PRINCIPLES IN THE QUANTITATIVE ANALYSIS OF WATERS, FERTILIZERS, PLANTS AND SOILS

DR. U.S. SREE RAMULU

Ph.D. (Calif), F.I.S.S.S., F.I.S.A.C., F.A.E.B., I.O.M.

Professor of Soil Science and Agricultural Chemistry (Retd.)

Tamil Nadu Agricultural University,

Coimbatore 641 003

SCIENTIFIC PUBLISHERS

Published by
SCIENTIFIC PUBLISHERS (INDIA)

Jodhpur –

5 A, New Pali Road
P.O. Box 91
Jodhpur - 342 001 INDIA

Delhi –

- 4806/24, Ansari Road
Daryaganj
New Delhi - 110 002 INDIA

Print 2018

ISBN : 978-81-72333-27-0

Visit the Scientific Publishers (India) website at
http://www.scientificpub.com

Printed in India

DEDICATED

TO THE ALMIGHTY, MY LORD

(for protecting me, providing me enough for happy living and above all for choosing me as *HIS TOOL to help HIS CHILDREN* at times of need)

AND

TO ALL MY TEACHERS

(from primary class to this day)

Dr. I.V. Subba Rao
Vice-Chancellor

Acharya N.G. Ranga Agricultural Ur
Rajendrai
Hyderabad - 500

FOREWORD

Books and laboratory manuals detailing methods of analysis of soils, water and plants have been written by different authors. A few of these books are either too simple leaving out essential details and principles of the analytical methods involved, while some other books, particularly those published by analytical chemists or by professional societies are of a very high standard. These are meant for research students and chemical analysts. But a good publication, including the basics and principles of the analytical methods involved and the methods of analysis needed for regular use by both UG and PG students and analytical chemists involved in soil, water fertilizer and plant analysis is much in need. This publication perhaps fills this gap.

The simple and lucid language in which the book is written should attract any reader whether he is a fresh analytical chemist or a trained one. To explain basics and principles, simple examples have been cited.

Dr. U.S. Sreeramulu with his varied experience in teaching and research, spanning over more than three decades definitely knows the needs of an analyst and some of the doubts which might be nagging them. The book tries to clarify these very well.

I am sure that the book which contains basics, principles and procedures for analysis of water, fertilizers and soils will be a useful publication which can be used as a Laboratory Manual also in addition to serving as a text on chemical analytical methods. The publication is particularly useful, for teaching courses in Soil Science, Water quality and Manure fertilizers. The students of Environmental Science and Analysts working in soil, water and Fertilizer testing laboratories can also make use of this publication as a ready to use manual. The book should find a place in all such laboratories.

I compliment Dr. U.S. Sreeramulu for having put in so much of his effort and time to bring out this very useful publication.

(I.V. SUBBA RAO)

FOREWORD

Soils, water and plant materials differ in their chemical composition and other properties. Physical, chemical and instrumental methods of analysis of these materials provide us an understanding of their nature, composition and properties and a base on which needed management interventions can be adopted for increasing the biomass and agricultural production on one hand and the quality of these resources on the other.

Standard methods of analysis with accurate determination need to be followed by analysts for prognostic purposes. It is said that 'The analysis can be no better than the sample'. Equally 'The interpretation and implication can be no better than the analysis'. It is in this context, detailed steps and procedures in the quantitative analysis of soils, plants, water and fertilizers, as lucidily brought out in this book by Prof. U.S. Sreeramulu is valuable to the large number of students in the academy and practicing analysis in numerous testing laboratories in the country, both in the public and private sector.

The thorough treatment of the subject in the book will, I hope, attract wide adoption and consultation.

M. Velayutham

CHENNAI

M. VELAYUTHAM
Former Assistant Director - General (Soils) &
Director, National Bureau of Soil Survey and
Land Use Planning

PREFACE

Rapid developments in the field of computer sciences and information technologies have completely changed the priorities of the younger generation as well as their parents and academicians. Further as these courses need very little infrastructure and also highly remunerative, many of the colleges are now offering these courses instead of basic science courses. Added to this, opening of large number of professional colleges in many states under self financing schemes have accelerated the changes further. These colleges are siphoning the cream of the students' population leaving only the rest for the conventional colleges. Even there the choice is more towards those courses which are remunerative in the immediate future. Even in the professional colleges due to the need to cater to the aspirations of the students as well as for financial gains, additional time is being allotted for the such courses at the cost of basic science courses. In some of the professional colleges, courses on basic sciences have been completely removed from the syllabus on the plea that the same is being taught in the high schools at higher level than is needed, not realising the fact that what is being taught now in schools is that of yesteryears and considerable amount of new knowledge is being added every day. All these have eroded into the basic understanding of the physical sciences and many who are forced to handle these courses now have not been exposed to the required extent to teach them properly, thus causing the fall in the standards.

These developments prompted me to provide a source material to all those who have to handle these subjects and that too at the level the students can understand. Hence nearly one third of the present book is devoted to the various principles that are adopted in the quantitative analysis of the various materials. Though in earlier years analysing the soils, plants etc., were done solely in a few select laboratories and that too for specific purposes, today due to various compulsions like the need to teach the pollution aspects etc., almost every college is offering some course or other related to environment wherein the analysis of these are to be done.

In contrast to other books on this subject wherein methods of analysis alone are given and that too starting with soils, then to plants etc. and finally to water, in this book the material is developed from the analysis of water which are similar to pure solutions, then to fertilizers which are similar to chemical compounds wherein one's analytical skill can be developed, and then from simple biological materials like manures to plants and finally to soils wherein considerable amount of preliminary treatments have to be done before the final chemical analysis. Further various sections have been written in such a way that those who have basic knowledge about chemical analysis or those who are not dealing with certain materials can skip those sections and proceed straight to the needed section and use it. Also due to various limitations that one has face in laboratories of the present day, only the methods that could be performed with the minimum glasswares and the common equipments have been described.

Almost every teacher from my primary school to my present teachers have been responsible for my developing interest in science. Apart from my chairmen *Drs. A. Mariakulandai, P.F. Pratt* and *N.T. Coleman* who have always been the guiding stars in my life, special mention has to be made of the following teachers who have motivated me towards science and for their special interest not only in my studies but also in my welfare; Messrs. *P.S. Yegneswaran* (Kotagiri Board High School), *N. Venkatakrishnan* and *T. Subramaniam* (Mannargudi Findlay High School) and *N. Ranganathachari, R. Subramainam* and *Dr. D. Raj* (Coimbatore Agricultural College). When books by Indian authors were rare and foreign books were beyond reach of any student, the text books on topics like heat, electricity, atomic physics etc., written in a simple language and suited to our needs and published my most beloved Physics professor Prof. M. Narayanamurthy of Vivekandanda College, Chennai, during my graduate classes have been the motivating force for my writing this book in such a way that even a student of average level can use it without difficulty.

There are a few persons/organisations who have laid the foundations for my professional growth and also in writing books for whom I am ever grateful. They are: Rockefeller Foundation, *Dr. S. Krishnamurthy, Dr. A. Venkataraman* and *Dr. M.S. Swaminathan.*

Normally in any book there will be only one foreword but in this book there are two forewords by two eminent soil scientists who have contributed much to the field of soil science and also who have helped me throughout. They are *Dr. I.V. Subba Rao,* (Vice Chancellor, ANGR Agricultural University, Hyderabad) and *Dr. M. Velayutham,* (formerly Director, Soil Survey and Land Use

Organisation, Nagpur and presently working as Coordinator in Dr. M.S. Swaminathan Foundation, Chennai). For both of them my sincere thanks are for ever.

The affection and encouragement showered on me by my parents (Late Sri U. Subbarayappa and Mrs. Subbalakshmi) and my brothers' families (Mrs. and Dr. U.S. Aswathnarayanan, and Mrs. and Dr. U.S. Krishnamurthy) can never be forgotten. Only the concerned can know and understand the value of the sacrifications made by the above people.

Also in the home front I was lucky to get the maximum cooperation from my wife Mrs. H.S. Rajamani and children Vasu, Sumathi, Seenu and Shoba. Added to this is the fun and frolic from my grandchildren Gouri, Sanju and Sadhana which relieved me of the monotony, depression and tiresomeness. I can never forget the contribution made by my granddaughter S. Gouri who put a lot of material into the computer, drew and xeroxed many of the diagrams for the manuscript, helped me in proof reading and also in many other ways in addition to relieving me of the monotony by her jokes and play. Also the help rendered by my daughter Dr. U.S. Shoba in checking some of the technical details is gratefully acknowledged. My sincere thanks are also due to Mr. D. Suryanarayanan for preparing the cover page layout. Much more than these, there is one person *Mr. C.S. Raghavan* who had in his own subtle way and with his method of encouragement, characteristic of his own, always made me work, complete writing books and also with sense of happiness. To this I should add the comments, advice and expectations of my chairman Dr. P.F. PRATT ("CONTINUE TO WORK") and that of Dr. A. Venkataraman ("LET YOUR GOOD WORK CONTINUE"), that of Dr. Usha ("WHAT NEXT AND WHEN CAN I EXPECT?") and that Dr. Homer D. Chapman (WHY NOT YOU DO ONE FOR OUR INDIAN CONDITIONS?). I will not disappoint them.

I will have to record my sincere appreciation to the publishers "Scientific Publishers (India) Jodhpur" for the amount of effort they had put in bringing this book in such a fine form and also in such a short time.

Finally I would like to express my sincere gratitude to all my readers and also request them to send their valuable constructive criticism.

COIMBATORE **U.S.SREERAMULU**

CONTENTS

SECTION – A

SECTION – B

APPENDICES

SECTION - A

CHAPTER – 1

INTRODUCTION

The vast explosion of knowledge in science that everyone experiences today is mostly due to the rapid advances made in the field of chemistry. These advances in turn have been possible due to the significant developments in analytical chemistry wherein the analyst aims to find out the nature of the constituents present in an unknown compound and then tries to quantify them for proper interpretation. The former study in analytical chemistry is called **Qualitative Analysis** while the latter is called **Quantitative Analysis**. The principles and techniques developed in each branch in the recent years have been so rapid and vast that each has now become a separate subject by itself.

Quantitative analysis in the earlier years was confined to analysis of pure compounds. Due to the growing needs of the present day and also due to the initiative of many analytical chemists, the principles and procedures developed for pure chemical compounds have been applied either as such or in a modified form for the analysis of many other compounds. For example in most of the quantitative analysis, the compound is first brought into solution by dissolving it in a solvent, the solvent in many cases being distilled water and then analysed. As water samples are just similar to the solutions described above, the same procedures have been adopted successfully for analysing water for drinking, industrial and agricultural uses. Similarly in the analysis of fertilizers and manures which are badly needed for increasing crop production in any country but at the same time available in limited quantities and hence frequently adulterated due to heavy demands, the procedures used for analysis of pure compounds have been adopted, as fertilizers are practically pure chemical compounds, almost meeting the analytical requirements of laboratory grade reagents. Slight modifications of these methods have enabled the analyst now to analyse manures which are complex organic compounds.

Use of waters, fertilizers and manures has been found to affect the productivity of soils on which they were applied and also the quality of produce produced in these soils. Hence a need arose to quantify the changes in the soils and plants before any valid interpretation could be made and also before any remedial measures could be taken. This has also been done by modifying the basic methods proposed for analysis of pure compounds in analytical chemistry. Though the methods developed upto now in these fields provide some basic information, much more refinement is still needed.

In quantitative analysis the methods of analysis have been grouped and classified differently based on one or more criteria. In one classification they are separated into gravimetric analysis, volumetric analysis and instrumental analysis. Briefly gravimetric analysis deals with analysis wherein a constituent or some of its compounds is weighed to find out the per cent of the constituent. Volumetric analysis measures the volume of a substance used to determine the per cent of a constituent. If the method involves a titration, it is called a titrimetric analysis. In the instrumental analysis, the constituent is estimated by measuring it with instruments based on optical properties (colorimetric methods) or electrical properties like conductance (conductometric methods), electrical potential or voltage (potentiometric methods), though these methods of analysis involve instruments which are costly. Still in view of the large number of samples that could be analysed with accuracy and also the complicated nature of most of the present day analysis, many laboratories are now equipped with a few of the instruments like pH meter, solu bridge, colorimeters, flame photometers, spectrophotometers etc.

In another classification, the methods are grouped on the purpose of analysis. If the compound is to be analysed for all its constituents, it is called "COMPLETE ANALYSIS". This is done only in factories manufacturing highly pure compounds conforming to certain accepted standards and also where they are required by law or otherwise to furnish the complete data. Normally in any laboratory only one or more constituents alone are estimated just like the determination of nitrogen, phosphorus, and/or potassium in fertilizers and bicarbonate, carbonate, chloride, sulphate and divalent ions in waters. This type of analysis is called "PARTIAL ANALYSIS". In certain types of analysis, one or more constituents in the sample which behave alike and are also extracted by the same treatment is determined either because it is not possible to estimate them separately or may not be worth estimating them individually as their influence is similar for certain studies in that system. For example iron and aluminium oxides in the hydrochloric acid extract of soil is determined as total sesquioxides using ammonia for precipitating both of them together as they behave similarly in fixation reactions in soils. This type of analysis is called "PROXIMATE ANALYSIS". In some other cases the per cent of each element in a substance is determined for establishing the formula of the compounds. In this analysis, the presence of trace elements is ignored, These analysis are classified under "ULTIMATE ANALYSIS".

Another classification is based on the size of the constituent present in the sample. If more than 0.1 gm of the constituent is expected to be present in the sample, the methods adopted for analysing them are called "MACROMETHODS". If the content ranges approximately from 0.01 to 0.1 gm, then the methods adopted come under "SEMIMICROMETHODS". The methods applied to estimate further lesser quantities come under "MICROMETHODS". The last one can be carried out only with instruments and the degree of sophistication of the instruments needed for the methods increase with the decrease in the size of the constituents. ULTRAMICROMETHODS of analysis are also adopted in samples of 1 to 10 microgram size (1 microgram = 0.001 mg) but they are not discussed in this book as they are not carried out on a routine basis in the analysis of waters, fertilizers, manures, plants and soils.

Vast interest in the above fields is being evinced in recent times due to their influence on the ecosystem and consequently on human life. This has resulted in this subject being taught even in high schools.

CHAPTER - 2

GRAVIMETRIC ANALYSIS

Gravimetric analysis or quantitative analysis by weight is an important technique in analytical chemistry. In this an element or a compound of definite composition of that element is isolated in a pure form from the substance and weighed. Usually the element or radical to be determined is converted into a pure, stable compound of accurately known composition and then estimated. The advantage of this type of analysis is that the constituent may be seen and examined for the presence of impurities and corrections applied for their presence. However most of the methods in gravimetric analysis are time consuming and laborious.

Isolation of the element or radical is done by precipitating the constituent into some stable insoluble compound of accurately known composition. Then its dry weight is found out after filtering, washing free of solution and impurities adhering to the substance, drying, igniting if necessary, and weighing. The accuracy of the methods depends upon the selection of the precipitating solution as well as the desired insoluble compound of the constituent in question. For accurate results, every trace of the precipitate should be collected and weighed. Under ideal conditions, gravimetric analysis is far more accurate than volumetric analysis.

In gravimetric analysis, in addition to the precipitation of the compounds, other methods like volatilisation or evolution of gas, electroanalytical methods and physical methods are also used. For example, volatile compounds such as water or carbon dioxide are often determined by volatilisation. In these separation of the constituent has to be sufficiently complete and the amount of the constituent that has been left out should be less than 0.1 mg. Generally the precipitation of any constituent is accompanied by complications such as formation of a colloidal solution, coprecipitation of impurities, post precipitation etc., and these introduce errors in the analysis which have to be minimised to the maximum extent possible.

Gravimetric analysis by precipitation involves number of steps like solution, precipitation, testing for completion of precipitation, washing, drying and igniting the precipitate to convert it to the required form. The first step in this analysis is to bring the given test compound into solution. This is done by weighing an exact quantity of the substance, transferring it to a clean beaker without any loss and then converting it into a solution by the addition of a solvent, generally water. In case the substance is soluble only in dilute acids, then the minimum required quantity of the acid is added carefully to bring it into solution. Wherever evolution of a gas is expected, then the beaker is covered with a watch glass while preparing the solution.

PRECIPITATION

The required element or radical in the solution is precipitated into an insoluble compound of known composition and then determined. Precipitation can be done in number of ways. Complete precipitation can be achieved by adding exactly equivalent amount of the precipitating reagent as required in the equation. But in practice it is not practicable to add the exact quantity. Hence a slight excess of the reagent is added slowly to a dilute and hot solution of the given weighed sample. In certain cases, the precipitate is warmed and/or allowed to stand with the mother liquor for sometime generally overnight. This "ageing" or "digestion" facilitates the easy coagulation of the precipitate to a filterable size.

The precipitate thus formed may be separated out and dried and weighed. For example, in the estimation of barium or sulphate, the element or radical is precipitated as barium sulphate ($BaSO_4$) and dried and ignited. During ignition the compound is not changed chemically. Similarly in the determination of silver or chloride, the element is precipitated as silver chloride (AgCl) and is weighed as such as it does not change chemically during drying.

In certain other estimations, the chemical composition

precipitate formed may be one and the composition of the weighed compound may be different. For example in the determination of iron or aluminium as hydroxides {$Fe(OH)_3$, $Al(OH)_3$} by adding ammonium hydroxide (NH_4OH) to their solution, when the precipitate is dried and ignited, the compounds get converted into their oxide form.

$$2Fe(OH)_3 \rightarrow Fe_2O_3 + 3H_2O$$

$$2Al(OH)_3 \rightarrow Al_2O_3 + 3H_2O$$

Similarly in the estimation of calcium, calcium is precipitated as calcium oxalate ($CaC_2O_4.H_2O$) and weighed as CaO

$$CaC_2O_4 . H_2O \rightarrow CaO + CO_2 + CO + H_2O$$

Such conversion to other forms becomes necessary as the precipitated form contains more than one species of the compound with different amounts of water of hydration as in the case of $Fe(OH)_3$ or with compounds of varying composition. But the first requirement of any gravimetric analysis is that the precipitate should contain only one species of the compound. Hence this compound has to be ignited so that the ignited form has the composition of one single compound. In such cases, the conversion into the stable forms should be done by simple techniques. The compound that is formed should be chemically stable for a fairly long time and also in the presence of water vapour or carbon dioxide present in the atmosphere or at higher temperatures. This can achieved by the proper choice of the final compound. For example calcium oxide precipitate formed by the ignition of calcium oxalate ($CaC_2O_4.H_2O$) readily absorbs water and carbon dioxide. Hence this is converted into $CaSO_4$ (calcium sulphate) by the addition of sulphuric acid in the crucible and later removing the excess acid by evaporation. In the case of silver chloride (AgCl) the precipitate undergoes partial reduction by the action of carbon and other products present due to the incomplete combustion of filter paper.

For better accuracy, the content of the element being determined in the precipitate should be as low as possible so that the errors in weighing, losses due to the solubility etc., can be minimised. During precipitation, there is the possibility of occlusion of various extraneous substances or ions from the solutions into the precipitate and these are not removed thoroughly by mere washing. In such cases, precipitates that are volatile and could be removed during ignition through volatilisation should be used for precipitation.

Only for this reason Fe^{3+} is recommended to be precipitated by ammonium hydroxide (NH_4OH) rather than potassium or sodium hydroxide (KOH or NaOH), silver (Ag^+) by hydrochloric acid (HCl) and not with sodium or potassium chloride (NaCl or KCl), barium by sulphuric acid (H_2SO_4) and not by sodium or potassium sulphate (Na_2SO_4 or K_2SO_4). However in certain cases, it may not be possible to overcome this problem through this technique as in the case of precipitation of calcium as calcium hydroxide {$Ca(OH)_2$}, wherein ammonium hydroxide (NH_4OH) cannot be used as the precipitate is soluble in excess of this reagent. In such cases either sodium or potassium hydroxide have necessarily to be used and at the same time, maximum care should be taken to wash the precipitate thoroughly.

In certain analysis, a particular ion has to be precipitated in the presence of other ions and at the same time, the precipitants commonly used precipitates the other ions also simultaneously. In such cases, reagent which specifically reacts with the required ions alone should be chosen. For example in the extracts of soil, when aluminium is precipitated as aluminium hydroxide {$Al(OH)_3$} with ammonium hydroxide, iron (Fe^{3+}) is also simultaneously precipitated as $Fe(OH)_3$. In such cases for precipitation of aluminium alone, sodium thiosulphate ($Na_2S_2O_3$) may be used, as the precipitate ($Al(OH)_3$ + S) could be filtered, washed and ignited.

$$2Al^{3+} + 3\,S_2O_3^{2-} + 3H_2O \rightarrow 2\,Al\,(OH)_3 + S + 3\,SO_2$$

The sulphur burns away and the $Al(OH)_3$ is converted into Al_2O_3. The sodium thiosulphate does not precipitate Fe^{3+} but merely reduces it to Fe^{2+}.

As all the precipitates are soluble in the wash solution or in the mother liquor to some degree or other, the accuracy of the determination varies with the degree of solubility. Hence proper choice of the precipitant is very important. Silver chloride is more soluble in 0.1N NaCl or 0.1N $NaNO_3$ than in water because of the formation of $AgCl_2$ in sodium chloride and $AgCl^+$ and $Ag_2NO_3^+$ in silver nitrate solution. Also high concentration of any salt should be avoided during precipitation. Almost all metal ions other than those of the alkaline elements readily form complexes of varying stability.

The pH of the solution significantly influences the slightly soluble salts such as those of relatively weak acids as phosphates, sulphates, carbonates and oxalates. In most of the estimations, it is difficult to find a specific precipitating solution that can mask all the interfering ions.

Hence for achieving complete precipitation and also a pure

stable precipitate of single species, the following points may be considered:

a. Proper precipitating solution should be selected for getting a pure precipitate with desirable properties.

b. The precipitating reagent should be added drop by drop so that the solution does not become supersaturated even from the very beginning.

c. During the addition of the precipitant, the solutions should be constantly stirred so that the crystals already formed come into contact with the supersaturated solution of the solid to enable it to deposit around them.

d. A hot or a solution at the same temperature of the precipitation should be added to the boiling solution, so that there will be maximum solubility at the time of precipitation and also the surface energy is maintained at the minimum.

e. As far as possible, precipitation should be carried out in dilute solution.

f. The precipitate that is formed, must be insoluble in the mother liquor as well as in the wash liquid.

g. The precipitating reagent should be added in slight excess only to complete the precipitation but not in great excess.

h. In cases of crystalline precipitate, keeping them overnight in contact with the mother liquor would decrease errors due to the coprecipitation and also give good readily filterable precipitates. But ageing or digestion has little effect on amorphous or gelatinous precipitates.

i. Where there is no risk of decomposition or initiation of side reactions, the precipitation may be carried out at higher temperatures as the velocity of crystallisation is increased leading to better formed crystals. Solubility is increased as the degree of supersaturation is decreased and conditions of coagulation is provided.

j. The precipitate should easily be separable from the solutions by filtration and should be easily washed free of soluble impurities. The size of the particles should be big enough to prevent their passage through the filter paper.

k. The precipitate formed should be washed with suitable dilute solutions of an electrolyte in which the precipitate has the least solubility and also formation of colloids is prevented.

l. If the precipitate is still contaminated due to coprecipitation or other causes, the precipitate may be redissolved in a suitable solvent and again reprecipitated.

m. The precipitate should be capable of being converted into granules or crystals of definite composition either by evaporation of the solvent or by ignition.

n. For carrying out the filtration easily and also for retaining all the precipitate on the filter paper, the particles of the precipitate should be fairly big in size. This can be achieved if the precipitate is allowed to stand at room temperature for 12 to 24 hours. In cases where the precipitate is not likely to decompose on heating, the precipitate along with the solution can be warmed for 45 to 60 minutes over a water bath under gentle heating. Boiling the solution should be avoided. This makes minute particles to go into solution and again get redeposited on the larger crystals, thus increasing the size of the other crystals and preventing their passage through the filter paper during the filtration. Also the impurities are prevented from getting adsorbed on or in the precipitate. This "self purification" helps in improving the purity of the precipitate and also avoiding coprecipitation of minute particles and impurities. Rapidly formed precipitates are irregular in shape and have larger area. Ageing helps in reducing the irregular surface to further deposition and dissolution and redeposition, causing reduction in surface and also adsorption.

TESTING FOR COMPLETION OF PRECIPITATION

Accurate values can be obtained only if the precipitation is complete. This is tested by allowing the precipitate to settle down completely and then adding a drop or two of the precipitating solution. If no turbidity develops, then some more of the precipitating reagent should be added and again tested for the completion of precipitation.

WASHING OF THE PRECIPITATE

The next step after testing for completion of precipitation is the separation of the precipitate from the supernatent liquid and also the adhering impurities from the surfaces of the precipitate. For this the supernatent liquid is decanted off and then the precipitate is washed or rinsed with a solvent or solution that removes everything that is nonvolatile from the crystals. It is preferable to use small quantities of wash liquid for each washing taking care to drain the solution completely than to use the whole quantity of the wash liquid in one lot. The impurities can be removed easily if the contents of the beaker is stirred after each addition, then allowed to settle and the supernatent liquid decanted.

Distilled water is generally used to wash the precipitates. In case the precipitate is not soluble in water, other solvents that do not react with the precipitate and are also volatile at the room temperature or at slightly higher temperature without decomposing the precipitate, may be used. For example, ammonium nitrate is preferred to ammonium chloride for washing ferric oxide. Generally an electrolyte having a common ion with the precipitate or a saturated solution of the precipitate is preferred as it reduces the solubility errors.

Washing should be done with the minimum quantities of wash solution. Excess washings will make the ions which have neutralised the charges earlier on the colloidal particles and caused the precipitate to coagulate to diffuse out and form colloidal particles again due to peptization. These colloidal particles because of their fine size will pass through the filter paper and cause significant analytical errors.

The crystalline precipitates can be dried at low temperatures or washed with solvents like alcohol which will evaporate rapidly, leaving the precipitate in a dry state. Curdy precipitates have greater surface area than the crystalline precipitates and hence they will adsorb greater amounts of impurities. However the crystals in curdy precipitates are loosely held together in porous curds and can be cleaned easily by washing. Generally solutions of strong electrolytes such as dilute nitric acid are used as their ions replace adsorbed ion by ion exchange and thus prevent peptization. The total surface area of gelatinous precipitates is very large due to the presence of very small crystals and also the charges on the particles are due to the pH of the solution. These gelatinous precipitates (for example hydrous oxide precipitate) are generally washed with electrolyte solutions such as ammonium nitrate.

DRYING

The precipitate, free from impurities should be separated from the filter paper and its dry weight found out. In most cases, the filter paper with the precipitate is kept in an air oven at 110-120□C to dry. After it has completely dried, the precipitate is separated from the filter paper carefully and weighed or the precipitate with the filter paper weighed and then the weight of the filter paper deducted from the total weight.

In certain cases, the precipitates may loose water, volatilise or become oxidised at temperatures less than even 100°C. In such cases, they are filtered in a filtering crucible and finally washed with alcohol followed by acetone or other solvents which will remove water and also volatilise at room temperatures or at lower temperatures than the decomposition temperature. Solvents which would volatilise at low pressures could also be used.

In contrast to the above, precipitates of hydrated aluminium oxide or silicon dioxide may have to be heated to as high temperatures as 1100°C to remove last traces of moisture.

IGNITION

In a few cases the precipitates have to be converted into compounds which are stable at room temperature, as in the case of calcium oxalate precipitate which is converted into calcium carbonate or calcium oxide and estimated. This is generally achieved by igniting the dry precipitate at a suitable temperature. Only precipitates which are not easily reduced by the carbon of the filter paper can be ignited. Precipitates of compounds like silver chloride cannot be ignited with filter paper. In such cases the precipitate is separated from the filter paper and both are ignited separately.

The precipitate with filter paper is transferred to a weighed crucible and gently heated to remove the moisture. Then the crucible is heated to higher temperature to char the paper. In case the filter paper or its gaseous products catch flame, the flame is extinguished by closing the lid of the crucible. After the paper is completely charred, the crucible is inclined and rotated over the flame to remove all the carbon by oxidation. The heating is continued for 10-15 minutes more and then cooled in a desiccator and weighed. In case there is some uncertainty regarding the completion of ignition process, as in the case of ignition of calcium oxalate to calcium oxide, the process of ignition is repeated till a constant weight is obtained. In case of compounds that absorb moisture on exposure to air, the crucible should be cooled in a desiccator, covered with the lid and weighed. Few examples of precipitated and weighed forms of the precipitates are given below:

Element	Form in which precipitated	Form in which weighed
Ag^+ or Cl^-	AgCl	AgCl
Ba or SO_4	$BaSO_4$	$BaSO_4$
I	AgI	AgI
Al^{3+}	Hydrous oxide	Al_2O_3
Fe^{3+}	Hydrous oxide	Fe_2O_3
Mg, P	$MgNH_4PO_4$	$Mg_2P_2O_7$
Zn	$ZnNH_4PO_4$	$Zn_2P_2O_7$ or ZnO

Inspite of all these, in gravimetric analysis errors may be introduced due to coprecipitation, post precipitation, solubility of the precipitates in solution etc., and many a time the cumulative effect of these errors may be significant. Hence they should be reduced to the minimum.

COPRECIPITATION

Coprecipitation is defined as the precipitation of any extraneous substance which is not generally precipitated under the given conditions by the precipitant. This causes serious errors in gravimetric analysis, particularly in solutions of low concentration. The contamination of the precipitate by substances which are normally soluble in the mother liquor is termed as coprecipitation. This occurs either through adsorption on the precipitates or through occlusion in the crystal lattice of the precipitates of other substances. As these contaminants are held very tightly to the precipitate, it is very difficult to remove them by washing. Coprecipitation is favoured by :

i. similarity in the structure of the coprecipitated compound and the precipitates

ii. similar ionic radii of both the compounds, the differences not exceeding 10^{-1} times of the replaced ions and the high solubility of the coprecipitated ions

Coprecipitation also can occur due to the adsorption of the contaminating ion on the rapidly formed, imperfect or very small crystals. Ions common to the precipitate are adsorbed more strongly. For example when barium sulphate is precipitated from a solution with an excess of Ba^{2+} ions, the barium ions get adsorbed strongly and give a positive charge to the particles. This in turn attracts the negative ions present in the solution and results in the rapid growth of the crystals with the adsorbed ions being trapped inside the crystal. Similarly when the precipitation is carried out in excess of sulphate ions the positive charges in the solution are attracted to the negative ions.

Degrees of adsorption vary with the nature of the precipitate and the ions present in the solution. Precipitates with greater surface area adsorb more materials than the bigger sized compounds. Also adsorption generally increases with decreasing dissociation or with increasing polarisability.

Coprecipitation by adsorption of soluble ions is not a serious problem with silver precipitates. Silver (Ag^+) and chloride (Cl^-) ions are adsorbed most strongly as the first particles of the precipitate to form. The adsorbed ions slowly diffuses into the solution as the end point is reached. Thus diffusion of silver from within the clumps of the crystals takes place. Therefore the solution should be shaken thoroughly for 30 to 60 seconds after the indicator changes colour to avoid a false end point. Silver ions diffuse more slowly than chloride.

When excess of sulphuric acid is added to a solution containing a mixture of $BaCl_2$ and $FeCl_3$, the main $BaSO_4$ precipitate is contaminated with $Fe_2(SO_4)_3$. This is observed as a brownish coloured precipitate on ignition.

$$Fe_2(SO_4)_3 \rightarrow Fe_2O_3 + 3\,SO_3$$

Coprecipitation errors could be minimised by reducing the amount of foreign substances in the solution by precipitating it in as a dilute solution as possible or by double precipitation or reprecipitation. Double precipitation can be used in the case of hydrous oxides but not for precipitating barium sulphate as no suitable solvent is available. Also the errors due to coprecipitation can be reduced by digestion or ageing as in the case of barium sulphate or silver chloride. Conversion of the contaminant into some other form could reduce the error as in the case of conversion of chromate contamination in barium sulphate to chromic acid. Also certain techniques like the order and speed of addition of reagents, temperature during precipitation, constant stirring while precipitating etc., can reduce the errors.

POST PRECIPITATION

When the solutions are allowed to stand for a long time, post precipitation occurs and this introduces errors in the analysis. Foreign compounds frequently get precipitated on the primary precipitate due to the delayed action and the quantity increases with time, high temperature and also upon whether the contaminant was added to the liquid after or before the formation of the primary precipitate. For example in the precipitation of calcium as calcium oxalate ($CaC_2O_4.H_2O$) in the presence of magnesium (Mg^{2+}), magnesium oxalate (MgC_2O_4) separates out gradually upon calcium oxalate which acts as a nucleus. The greater the time of contact, the greater will be the amount precipitating on it and hence greater will be the error. Similarly zinc sulphide (ZnS) is post precipitated during the precipitation of CuS or HgS in 0.3 N hydrochloric acid in the presence of zinc ions. In all such cases, if the filtration is done within a short time after the completion of precipitation and in certain cases after adjusting the pH, the errors could be minimised. The important differences between coprecipitation and post precipitation are:

	Coprecipitation	Post precipitation
i	Amount of contamination decreases with time	Increases with time
ii	Contamination decreases with the speed of agitation of the solution	Increases with speed
iii	Magnitude of contamination is much less	Is greater

SOLUBILITY OF THE PRECIPITATES

Significant error could be introduced due to the solubility of the precipitate in excess of the precipitating reagents and also wash solution. In gravimetric analysis, though theoretically only the equivalent quantity of the precipitating solution as required in the equation is expected to be added, still in practice, out of necessity, a slight excess of the reagent has to be added to get complete precipitation. This affects the equilibrium between the precipitate and the ions. The larger the excess of the reagent the greater is likely to be the error. Hence the addition of large excess should be avoided so that the errors due to this could be minimised.

The solubility of the precipitate in the wash solution also introduces errors. Every precipitate is soluble to some degree in the wash solution and hence the least number of washings should be given to reduce the error due to this factor. Though from theoretical considerations, the quantum of error can be calculated, still it is not likely to be accurate as variations in time of contact during filtration after each addition of wash solution occurs and also the time of contact normally allowed in most cases are not sufficient to get the calculated amount in solution. Hence wherever possible, the wash solution in which the precipitate is least soluble should be selected. Presence of organic solvents such as methyl, ethyl, amyl alcohols, acetone and ether reduce the solubility of ionic salts. For example magnesium oxalate is moderately soluble in water but very slightly soluble in acetic acid solution. In cases where the solubility of a precipitate is appreciable in the common wash liquids, then a saturated solution of the precipitate can be used as it will reduce the losses due to washing.

The pH of the solution in some cases significantly influences the solubility as in the case of slightly soluble salts of relatively weak acids such as phosphates, sulphates, carbonates and oxalates. For example barium sulphate is ten times more soluble in 1 N acid than in water wherein its solubility is only 1 x 10^{-5} molar. As in the above case, slight change in pH can also be beneficial. Little acidity increases the solubility of phosphates, carbonates, and other possible contaminants and also improves the crystal growth by raising slightly the solubility of barium sulphate. Large crystals are more easily formed from solutions of high rather than low solubility. Also the presence of acids and bases increase the solubility of many of the precipitates. For example magnesium ammonium phosphate ($MgNH_4PO_4$) is soluble in acid while zinc ammonium phosphate ($ZnNH_4PO_4$) is soluble in excess ammonia and acids.

Higher temperatures of the solution or wash solution increases the rate of solubility of the precipitate and thus increase the error. For example barium chloride ($BaCl_2$) has a solubility of 0.00015 gm/100 ml of water at 20□C and 0.0022 gm/100 ml of water at 100□C. Hence while precipitating and washing the precipitate, conditions which favour least solubility of the precipitate in the mother liquor as well as in the wash liquids should be selected so that the errors due to these could be minimised.

In the analysis of materials discussed in this book only estimation of sulphate is done by gravimetric methods of analysis and that too only in a few laboratories. In all other analysis, either volumetric or instrumental methods are adopted. Hence they are dealt in much detail in subsequent pages.

CHAPTER - 3

VOLUMETRIC ANALYSIS

Volumetric analysis is an important branch in quantitative analysis. It consists essentially in determining the volume of a solution of accurately known concentration required to react quantitatively with the solution of the test substance. The volumetric analysis is also called "TITRIMETRIC ANALYSIS" as the techniques of estimation involves the process of titration. Volumetric analysis as a rule is capable of high accuracy (one part in 1000) and also possesses number of advantages over gravimetric methods. These analysis can be done with simple apparatus and also rapidly. Once the required reagents have been prepared, the titration of an individual sample can be done in a matter of minutes. A skilled technician can carry out hundreds of estimations in a single day. Hence volumetric analysis is adopted for large number of studies than other techniques.

Volumetric analysis depends on the measurements of volume of solutions. The amount of a substance in a sample is determined indirectly by measuring the volume of a solution of a known concentration (called the standard solution) required for the reaction with the desired constituent or with another solution chemically equivalent to the desired constituent. Generally the standard solution is taken in a burette and is allowed to mix gradually with the test solution until chemically equivalent quantities of the two have been brought together. To find the stage at which chemically equivalent quantities have been added, the analyst depends upon some change in the physical properties like change in colour, appearance or disappearance of turbidity etc. In most of the titrations a substance called the INDICATOR is added to indicate the end point. At the end point the indicator changes colour. The end point has to be as close as possible to equivalence point (called the stoichiometric end point). The entire process is called "TITRATION"

Accuracy of the analysis depends upon the standard solution used. A standard solution is one wherein a known weight of a substance is dissolved in an exact quantity of the solvent, generally water. The number of grams of substance per milliliter of the solution is known as the concentration or the titre of the solution. In certain cases, the substances can be weighed accurately but may not give a solution of the required concentration due to number of factors like instability in solvent, decomposition due to exposure to light, or due to reaction with organic matter etc. In the case of few other substances, they cannot be weighed accurately but once dissolved in solvent they give a stable solution. But an ideal standard solution is one whose substance can be weighed accurately and also is capable of forming stable solutions which come up very near to the above requirements and these are called PRIMARY STANDARDS and are used for the preparation of standard solutions. Sometimes the primary standards themselves cannot be used for certain titrations. In such cases the other solution is standardised with the primary standards. These standards thus prepared are called SECONDARY STANDARDS.

For preparing a standard solution, the required quantity of the substance is weighed accurately, dissolved in a measuring flask with water and the volume is made up to the mark with water. If "W" grams of the substance is dissolved in "V" ml of the solvent (or water) then the titer is T = W/Vgrams/ml. Often this method is not applicable in the case of substances like hydrochloric acid (HCl), sodium hydroxide (NaOH) as the exact concentration of an aqueous HCl is never known. Even if an exact weight of such a solution is taken, it is impossible to calculate the weight of HCl it contains. In the case of sodium hydroxide, it absorbs carbon dioxide and water vapour from the atmosphere and this results in a change in the weight quickly. In such cases the standard solutions of these are prepared by standardising them with other primary standards.

Different types of chemical reactions like neutralisation of hydrogen or hydroxide ions in the analysis of acids or bases, oxidation of ferrous ion in the determination of iron, precipitation of silver chloride in the determination of silver or chloride ions, formation of chelate complex in the estimation of calcium content of water are used in the volumetric analysis. Almost all types of reactions adopted in the volumetric analysis of the materials are discussed in this book.

STANDARD SOLUTIONS

Accuracy in volumetric analysis depends upon the use of a solution of known concentration for determining quantitatively the amount of constituent present in the test solution. The solution whose exact concentration is known is called the STANDARD SOLUTION. This solution can be prepared in two ways.

Direct method: In the direct method, a solute of definite and known composition is dissolved in a solvent and made upto a known volume in a volumetric flask. Generally water is used as solvent. The solute should be of very high purity and should have the characteristics of a primary standard. This method is most suitable for preparing a standard solution of predetermined concentration such as 0.1 N or a solution of an exact titer of the specified constituent eg. a calcium titer of 5.00 mg/ml.

Indirect method: In most of the cases, solutes to be used for the preparation of standard solutions do not conform to the requirements of a primary standard. Hence a solution of an approximate concentration is made up after careful weighing and transferring the same to the volumetric flask. This solution is then standardised by measuring accurately the volume of the solution required to titrate an exactly known quantity of a primary standard. The actual concentration is then calculated from the volume of primary standard used adopting the usual procedure.

Both these methods require the use of a primary standard which should react stoichiometrically i.e. react exactly as the equation indicates. A solute to be used as primary standard should have the following characteristics:

i. *Purity* : The solute should be of the highest possible purity, preferably 100%. Normally it may not be possible to get solutes of this purity but those of lesser but exactly known purity can be used. Also it should be possible to test for the amount of impurities. Solutes with information on the percentage of purity and also on the nature and amounts of impurities present can be used with advantage.

ii. *Stability* : The substance should be solid and should not undergo any chemical change in storage even kept for a long time. It should not absorb water or carbon dioxide or undergo oxidation when exposed to atmosphere. It should be stable at the drying temperature used to drive away absorbed water or water of crystallisation (approximately 110-120□C). In order to ensure complete freedom from occluded moisture, the material can be fused without decomposition. In case of liquids, loss by evaporation should be prevented by weighing the material in a closed vessel and also transferring the same quickly to the volumetric flask. It is advisable to discard a few grams of the substance found on the top surface of a bottle before taking the material for weighing.

iii. *Chemical properties* :

a. The chemical should react quickly and also stoichiometrically with the solution that is being standardised.

b. *High equivalent weight*. A primary standard should have a high equivalent weight but it is not absolutely necessary. The higher the equivalent weight the larger will be the amount of material needed for preparing the standard solution and lower will be the relative error due to weighing.

iv. *Availability* : The reagent intended for primary standard should be commonly available in any place and also should not be very costly. It should be available preferably in solid form.

Hence a reagent to be used as primary standard should possess the following characteristics:

i. highest possible purity,

ii. their exact composition and also the quantum of impurities known,

iii. could be weighed accurately within a reasonable time and with the least precautions,

iv. should not be hygroscopic,

v. should be capable of being dried without decomposition at atleast 110□C, cooled in a desiccator and

vi. should react stoichiometrically and also quickly.

Preparation of standard solutions

The material for a primary standard should be dried in an air oven at 100-110□C or more to remove the moisture and then cooled in a desiccator. The required quantity of the reagent should be weighed accurately and transferred to a volumetric flask without loss. Then a small quantity of the solvent should be added and swirled well so that all the materials are completely dissolved. Or alternatively the weighed material is transferred to a 200 ml beaker, a small volume of solvent is added to dissolve the substance completely and then the solution is transferred to a volumetric flask. The beaker is then rinsed with small aliquots of the solvent and the washings added to the volumetric flask. Then the same is made upto the volume and preferably kept overnight. The resulting solution will have a known concentration because the weight of the reagent added to a given volume is exactly known. In certain cases, relatively large amounts of expensive primary standard material may have to be used and in such cases only a small volume of the primary standard should be prepared.

The same primary standard cannot be used for every purpose. Different types of primary standards are prepared to suit each purpose. The following are some of the primary standards used in volumetric analysis:

i. *Primary standard acid :* The primary standard acid is used for standardising a solution of the base. Common primary standard acids are:

   a. Potassium acid phthalate ($KHC_8H_4O_4$; Molecular weight 204.23; equivalent weight 204.23). It is monobasic acid, has high equivalent weight, is anhydrous, stable, nonhygroscopic and hence useful for the standardisation of strong carbonate free bases using phenolphthalein as indicator. However it is not suitable for titration with ammonium hydroxide.

   b. Benzoic acid (C_6H_5COOH; Molecular weight 122.13; equivalent weight 122.13). This has also got high equivalent weight but is insoluble in water. It must be dissolved in water-alcohol solution.

   c. Sulfamic acid (HNH_2SO_3. Molecular weight 97.11; equivalent weight 97.11) This has also high equivalent weight and is also soluble in water.

   d. Oxalic acid ($H_2C_2O_4.2H_20$; Molecular weight 126.06; equivalent weight 63.03). This is also a satisfactory primary standard. However because of the uncertainty in the number of molecules of water of crystallisation, in the dry climate the finely powdered oxalate crystals may absorb or have different amounts of water of crystallisation. Hence these have to be dried and weighed quickly without much exposure to the atmosphere particularly in humid climates. Decinormal and less concentrated solutions are unstable and should be prepared fresh everytime. Higher concentration solutions are fairly stable at room temperature when protected from light.

ii. *Primary standard bases* : These are generally solid bases used for standardisation of acids.

a. Sodium carbonate (Na_2CO_3; Molecular weight - 106.3; equivalent weight 53.15). This is the commonest primary standard base used for standardisation of the solutions. One of the disadvantages is that it may pick up carbon dioxide from the air. The bicarbonate that is formed may be decomposed by heating it to 270□C since carbonate ion is a weak base and bicarbonate is still weaker. This reagent can be used only for standardisation of strong acids such as hydrochloric acid or nitric acid. Methyl orange and methyl purple can be used as indicator. Pure sodium carbonate may also be prepared by heating a known weight of pure sodium bicarbonate on a sand bath till constant weight is observed after cooling in a desiccator.

b. Calcium carbonate ($CaCO_3$; Molecular weight 100.0; equivalent weight 50.0). This could be used in the place of sodium carbonate and the reactions of both substances are identical.

c. Borax (Sodium tetraborate; $Na_2B_4O_7.10H_2O$; Molecular weight 381.44; equivalent weight 190.72). Borax is recrystallised and forms a water solution and is equilibrated with sodium bromide in a desiccator before using it as primary standard. The difficulty of preparing it however prevents it use except in rare cases when no other compound is satisfactory.

iii. *Primary standard oxidising agents*

a. Potassium dichromate ($K_2Cr_2O_7$; Molecular weight 294.21; equivalent weight 49.04). This is easily purified and easily dried in an oven at about 140□C. It is stable and inexpensive enough to use for making up solutions directly.

iv. *Primary standard reducing agent*

a. Sodium oxalate ($Na_2C_2O_4$; Molecular weight - 134.0; equivalent weight 67.0). This is used in the standardisation of potassium permanganate and cerate solutions. It can dried safely at 105□C but must not be heated along with acid solutions to near boiling as it decomposes forming oxalic acid in sulphuric acid solution. This reagent is available in very pure form and is also stable in dry condition.

b. Arsenic trioxide (As_2O_3. Molecular weight - 197.82; equivalent weight 49.435). This is used to standardise iodine, permanganate and cerate solutions. As it is difficult to dissolve it in water, a pellet of sodium hydroxide with a little of water can be used. Sodium bicarbonate is used in large excess to buffer the solutions during the titrations.

c. Electrolytic iron (Fe; Atomic weight 55.85). This is used to standardise dichromate, permanganate and cerate solutions. This is available in a high degree of purity. The iron is dissolved in hydrochloric acid and the ferric iron is reduced to ferrous form with stannous chloride and zinc and the titration is carried out in acid solution containing phosphate. The phosphate is added to form a stable colourless compound with ferrous ions formed during the titration. Then ferric ions' colour does not interfere with the colour changes at the end point.

v. *Other primary standards* :

a. Sodium chloride (NaCl; Molecular weight - 58.454; equivalent weight - 58.454). This is used in standardisation of silver nitrate solutions that are employed in the estimation of chloride ions.

b. Constant boiling hydrochloric acid: This is sometimes used as a primary standard. Many of the handbooks on chemistry will give tables relating the concentration, specific gravity and other properties of constant boiling solutions of this acid. These may be referred to whenever needed.

c. Succinic acid ($COOH(CH_2)_2COOH$. Molecular weight 119.0; equivalent weight 59.50). This is a dibasic acid and can be dried at 105□C. This is a stable compound and can be used for preparing decinormal primary standards.

SECONDARY STANDARDS

Certain reagents may lack one or more of the desirable characteristics of a primary standard but still be very useful for preparing a standard solution. They are known as "Secondary Standards" and they are commonly prepared in bulk for routine analysis while the primary standards are prepared in small quantities and that too in many cases for standardisation purposes only.

In volumetric analysis, it is frequently necessary to prepare a definite volume of a standard solution of required strength, or a solution of a known volume of the strength approximately close to the required strength is prepared and then standardised against a primary standard. Normally solutions of a normalities like 0.1, 0.5, 1.0 N are prepared for ease of calculations.

The following guidelines may be followed for preparing a solution of the required strength;

a. The quantity of the reagent needed to prepare the solutions of the required strength may be calculated from its equivalent weight.

b. Then slightly higher amount of the reagent is weighed and dissolved in the desired volume of the solvent. This is to provide an allowance for any possible decomposition reactions, for the presence of impurities and also for diluting the solution to the required strength after titrating it with the primary standard.

c. Then a known aliquot of the prepared solution should be taken and titrated against the primary standard using appropriate indicator. From this, the exact strength of the prepared solution is calculated.

d. Based on the strength of the prepared solution, the volume of the solution needed to prepare the required volume of the solution of given strength is also calculated after deducting for the amount already used.

e. Then the calculated quantity of the prepared solution should be transferred to a container with the help of a pipette or burette and made upto the mark with the solvent. After shaking it well, it may be standardised against the primary standard. It is preferable to keep the prepared solutions overnight observing all the precautions for storing of standard solutions. Next day it may be standardised again before using it, to ensure that its concentration is "as required".

f. *Example* : Preparation of decinormal potassium permanganate

i. The molecular weight of potassium permanganate is 158.03. The equivalent weight is 1/5 th of the molecular weight as per the equation given below:

$$2\,KMnO_4 + 3H_2SO_4 \rightarrow K_2SO_4 + 2\,MnSO_4 + 3H_2O + 5\,(O)$$

As two molecules of $KMnO_4$ is equivalent to 5(O), the equivalent weight of potassium permanganate is 158.03/5 = 31.606 which indicates that 31.606 grams of the substance should be dissolved in one liter of the solvent to give 1N solution or 3.1606 grams should be dissolved in the same volume to give 0.1 N solution.

Normally slightly higher amount of potassium permanganate is weighed and a solution of approximately higher strength than that is needed is prepared by dissolving it in the required amount of water. As distilled water always contains some amount of organic matter, it slowly oxidises permanganate, thus resulting in a change in the strength of the solution. Hence the permanganate solutions are generally kept overnight to allow for these reactions to be completed so that there may not be further change in the concentrations due to these reactions later. Also care should be taken to see that the solution does not come into contact with rubber, filter paper or any organic material.

For standardisation, the permanganate solution is taken in a burette after observing all the precautions and titrated against a known amount of hot standard oxalic acid solution containing dilute sulphuric acid. For the standardisation usually 25 ml of decinormal (0.1N) oxalic acid is used. This standard oxalic acid solution should be heated slowly before titration but not boiled. Then the hot solution is titrated against permanganate with constant stirring. The permanganate that is added will get decolourised as soon as it comes in contact with the oxalic acid in the presence of sulphuric acid;

$$2KMnO_4 + 5H_2C_2O_4 + 3H_2SO_4 \rightarrow K_2SO_4 + 2MnSO_4 + 10CO_2 + 8H_2O$$

Titration should be continued till the appearance of the pink colour persists for some time even on shaking. Then the volume of the permanganate solution used is recorded and using the formula $N_1V_1 = N_2V_2$ (wherein N_1 = Normality of oxalic acid used; V_1 = volume of oxalic acid taken; N_2 = normality of permanganate solution = not known; and V_2 = volume of permanganate solution used - this is given by the titer value), the strength of the permanganate solution is determined. Since three out of four factors are known, the fourth factor viz. the concentration of permanganate solution can be calculated.

Using the same formula, the volume of permanganate (approximately 0.1 N but exact value determined by the previous estimation) required to prepare a required amount of 0.1 N standard permanganate solution can be calculated. N_1 = normality of the solution needed (0.1N); V_1 = volume of 0.1 N standard solution needed (say 1000 ml); N_2 = normality of the solution already prepared (say 0.102 N). V_2 = volume of this solution to be taken (to be determined).

$N_1V_1 = N_2V_2$ 0.1 x 1000 = 0.102 x Y.

From this Y (the volume of already prepared solution to be taken) is calculated. Then take this amount (Y ml) in a 1000 ml volumetric flask and add water to make up to the volume. This will give 1000 of 0.1 N potassium permanganate. The formula $N_1V_1 = N_2V_2$ is an important formula which is used in every titration and hence should be remembered.

In volumetric analysis, the concentration of solutions are expressed in number of ways depending upon the purpose. The following are some of common terms used in expressing as well as in calculating the concentrations of the various solutions :

MOLAR (M)

A molar solution is one which contains one gram molecular weight or one gram - formula weight of a substance dissolved in one liter of a solution. A solution may have the same molarity as normality if the equivalent and molecular weights are equal.

FORMAL (F)

The term "formal" is applied to ionic species while the term "molar" is applied to either ionic or molecular substances. A formal solution is one which contains one gram formula weight of the substance in one liter of the solvent. The substance in solution may be in several ionic forms or species. For example 0.1 formal solution of sulphate ions may be partly bisulphate ions or even partly sulphuric acid but the solution is said to be 0.1 F (formal) with respect to sulphate.

MOLAL (m)

A molal solution is one where one gram formula weight or one gram molecular weight of the solute is dissolved in 1000 gm of the solvent.

NORMAL (N)

A normal solution contains one gram equivalent weight of the substance per liter of the solution. An equivalent weight is the amount of the material that will react with one atomic weight of hydrogen or its chemical equivalent. If the equivalent weight and the formula weight are the same, then the solution will have the same normality and molarity.

PER CENT BY WEIGHT (%) (w/w)

Per cent means parts per 100 parts. Generally a solution when expressed as per cent is assumed to be per cent by weight. In this solution an known weight of the substance in grams is dissolved in 100 grams of the solvent or in 100 ml of water. If the units of weight are different, (lb or kg) then the same definition holds with slight modification, that is, instead of grams the word pound or kilogram is substituted.

PER CENT BY VOLUME (%) (w/v)

The definition is similar to the previous one except for the fact that the strength of the solution is expressed on volume basis. In this a known weight of the substance expressed in grams is dissolved in 100 ml of the solvent.

TITER

The titer of a solution is the weight of a substance equivalent to one ml of the standard solution. For example the milligram of chloride equivalent to 1 ml of silver nitrate solution is called the chloride titer of silver nitrate solution. Whenever the titer is written, the units of weight per unit of volume should also be indicated.

Titer is very useful in quickly calculating the results of titrations. The titer is the weight of any constituent in a substance chemically equivalent to one ml of the solution used for titrating the constituent and is numerically equal to the normality of the solution times the "gme" of the constituent sought. It reduces the number of factors in the equation.

$$\text{Per cent of the constituent} = \frac{V \times N \times gme \times 100}{\text{Weight of the sample}}$$

GRAM MILLIEQUIVALENT (GME)

The gram milliequivalent is that 1 ml of 1 N solution contains 1 gram milliequivalent of the reactant. When a reactant is weighed, dissolved and titrated, calculations can be best made by using gram-milliequivalent (gme). The value is gram equivalent weight of the reactant divided by 1000.

$$\text{gme} = \text{Gram} - \text{equivalent weight}/1000$$

Since the volume times normality of any solution equals milliliters of 1 N solution equivalent to the volume, the number "gme" (a weight value) and ml of 1 N solution (a volume value) are always equal to each other. Thus "gme" is a value common to both weight and volume.

CHAPTER - 4

GENERAL PRINCIPLES IN VOLUMETRIC ANALYSIS

Volumetric (titrimetric) analysis is one commonly preferred for the estimation of macroquantities of any compound as it is less time-consuming, easy to perform and also requires least number of apparatus. At the same time it involves number of steps from preparation of the standard solutions to the final titration of the unknown solution. Even small errors committed in each step will add up and be of significant magnitude when their cumulative effect is considered. Hence it is absolutely necessary to minimise the errors to the maximum extent possible. The following important guidelines would help in increasing the accuracy of the final result.

The sample taken for analysis or the analytical grade reagent taken for the preparation of the standard solutions should not be too small. If very small quantities are taken, then significant amount of errors are likely to be introduced during weighing, transferring, making up of the solutions etc. These errors will further get magnified if the content of the element in the compound or the equivalent weight of the standard solution is low.

Also the aliquots of the unknown solution or the standard solution should not also be too small. Errors due to improper delivery into the containers from pipette, calibration errors of the apparatus, materials sticking to the sides of burette or pipette and not transferred to the containers will be proportionately large in smaller aliquots and when multiplied by a factor of dilution to convert it to the original volume or weight, the cumulative error will be very large.

Invariably many analysts attempt to complete the titrations within 10 or 15 ml under the wrong notion of reducing the cost per estimation and also carrying out more analysis with the available standard solution, thus avoiding the preparation of additional amount of new standard solutions. Also they feel that the burettes need not be filled frequently. In any titration, burette reading error of 0.02 ml and a drainage error of 0.02 ml making a total of 0.04 ml can always be expected. Added to it the accuracy of calibrations of the volumetric apparatus particularly those supplied commonly nowadays without certificates require to be much desired. Though the total volume of 25 ml or 50 ml delivered by the burettes may be indicated on the burette, still when the actual volume delivered between 0 to 10, 10 to 20, 20 to 30, 30 to 40 and 40 to 50 ml is measured, they may not exactly be 10.0 ml in each case. The error due to this can be reduced by using larger volumes of the titrants. Wherever possible it is better to use certified volumetric apparatus but they are costly and may pose number of problems later when they have to be condemned or disposed of due to wear and tear. Hence the available apparatus may be checked with the help of a certified apparatus and only those conforming to the standards may be used in the analysis wherein greater accuracy is expected. The volumetric error from all sources of errors should be kept below 1%.

At the same time the sample or aliquot taken for analysis cannot be increased indefinitely beyond a certain level as this will result filling the burette number of times even for a single titration. This is not only cumbersome but would also introduce further errors like reading errors, drainage errors, errors due to lack of sufficient intensity of colour near the end point etc. Hence a proper choice of the quantity of the sample and aliquot has to be made such that the sample is neither too large nor too small.

Each one of the titrations should be carried out continuously and completed without break. Any discontinuity in the titration invariably increases the error component due to contamination from other sources, side reactions, reactions with atmospheric gases and moisture from the atmosphere.

The reagents used in the preparation of the standard solutions and other reagents should be as pure as possible, preferably of Analar quality and also fresh. On number of occasions, solutions are prepared from reagent remaining in the bottle that has been opened and used earlier. It is not uncommon to find many of the reagent bottles not closed at all with the lid for number of days or not closed properly. Hence it is better to use fresh reagents or in case of costly reagents, the top layer of the reagent should be scrapped carefully and discarded and the lower part of the reagent alone used for the preparation of the solutions and the solutions should be used only after proper checking and standardisation.

The reagents and the prepared standard solutions should be stored in proper containers and also kept at low temperatures. Hydroxides of sodium or potassium as well as other strong alkalies should be stored in plastic containers or in a glass bottle waxed inside to prevent their action on glass. Care should also be taken to prevent absorption of carbon dioxide from the air, usually done by filtering the air through absorbents like sodalime. Certain reagents like silver nitrate should be stored in coloured bottles and in the dark to prevent any photodecomposition of the salt. Also solutions like potassium permanganate should not be allowed to come in contact with organic materials like rubber cork, filter paper or organic matter. These steps will reduce the errors due to the instability of the reagents in storage.

After titration is over, there is always a tendency to return the standard solutions remaining in the burette to the original containers with the intention of saving the costly reagent and also carrying out more number of analysis with the amount of solution already prepared. But in reality, this is a faulty notion as it results in transferring the contaminated or diluted solution to the original pure standardised solution and wasting the whole lot as irrespective of all the care taken by the analyst, some dilution or contamination always occurs in the burette and during titrations and this cannot be avoided.

The accuracy of the final results depends upon the choice of the concentrations of the standard solution and the test solutions. Generally standard solutions of 0.05 N to 0.2 N are preferred with 0.1 N being widely accepted. Stronger solutions give sharper end points but even a small error or a drop, that is 0.05 ml in 10 ml would introduce an error of 1/200 or 0.5%. So dilute standard solutions are preferred but even here there is lower limit to which one can dilute the solution. In very dilute solutions, errors due to the lack of sharpness in the end point eventually become larger than the errors due to burette readings and other sources. Also the rate of reaction in dilute solution may be too slow to perceive. Hence the concentration of the standard solution should be so chosen that the volume in milliliter read on the burette is per cent of the desired constituent. For this the concentration of the standard solution is so adjusted that one ml of the standard solution is equivalent to 0.01 g of the weight of the sample assuming that the sample is 100% pure.

Greater accuracy in titrimetric analysis can be achieved if the reaction selected is such that only one single reaction occurs between the desired constituent and the titrant. At the same time the reaction should occur as indicated by the equations and that too rapidly, so that only a little time is consumed in carrying out the titration. Fortunately most of the ionic reactions are so rapid that they can be considered as instantaneous. The other reactions which are not so rapid can be made to proceed faster by the addition of a catalyst. Also the reactions must be completed when the equivalent amounts of the reactants are brought together.

In choosing any particular reaction for the titrimetric analysis, it should be seen that a suitable standard solution and an indicator having the desired properties are also available. Also the accuracy of the estimations can be increased if loss of sample or solution even inadvertently during weighing, transfer etc., is avoided and also proper order of mixing the reagents are followed during the estimations.

CLASSIFICATION OF ERRORS IN ESTIMATIONS

Even in fairly uniform sample where number of measurements are made for the same parameter by the same technique and by the same analyst, the results of each estimation is likely to differ from each other slightly or widely due to some errors known and unknown, determinable or undeterminable, that could be rectified and eliminated or not rectifiable at all. Though the analyst always tries to eliminate all possible errors, he may not be able to eliminate them completely in all cases. But he can certainly minimise their effects to the maximum possible extent by observing certain precautions. The analyst working with the best equipment and using the most accurate technique available for analysis as on date may get different values for the same parameter particularly in biological systems. For practical purposes the average value is to be taken as the most acceptable one. The degree of error and sources of variation are classified as follows;

i. *ABSOLUTE ERROR* : This is the difference between the observed or measured value and the true value of the quantity measured. In biological systems it is very difficult to know the true value of any parameter. Hence the absolute error can be taken as the difference between the observed or measured values and the most probable value of the quantity measured. An analysis is said to be performed accurately if these two values agree with each other.

ii. *RELATIVE ERROR* : This gives a measure of the degree of error. It is the ratio of the absolute error divided by the true or most probable value. In accurate analysis, where both values agree, the relative error is zero. Usually the relative error is expressed in per cent or parts per thousand.

The various types of errors encountered in the analysis may be grouped into two categories : i. Those that can be determined and eliminated or reduced to the minimum by following certain techniques. These are called DETERMINATE OR CONSTANT ERRORS. ii. The others are unknown errors but definitely occur in the analysis. The actual contribution or source of these errors cannot be determined by any of the technical methods available on date. These are called INDETERMINATE ERRORS or RANDOM ERRORS or ACCIDENTAL ERRORS.

i. *Determinate error* : The sources and magnitude of these errors can be determined and they can be either avoided or their contribution minimised. Most of the errors are due to operational and personnel errors for which the analyst is responsible and not the method or technique. The determinate errors can be further grouped as follows; a. Physical errors, b. Personnel errors, c. errors in the choice of methods and d. additive or proportionate errors.

i. *Physical errors* : The cause for the physical errors can be any one or more of the following :

a. Loss of materials in various stages of analysis.

b. Under or excessive washing of the precipitate.

c. Use of wrong type of filter paper or igniting the precipitate at the incorrect temperature or allowing the by-products of ignited paper to react with the precipitate.

d. While dissolving the precipitate for further titration, adding the filter paper at the wrong stage or pulping the paper when it should not be done.

e. Slowly transferring the crucible to the balance from the desiccator and sometimes keeping the desiccator open while weighing other samples and also for weighing the sample. The magnitude of error due to this will be more particularly when dealing with precipitates which are of hygroscopic nature.

f. Use of reagents of poor quality, the impurities in such reagents may also react with the titrating solutions or add to the weight of the precipitate or delay the occurrence of the end point.

ii. *Personnel errors*: These errors are due to the analyst's inability to make certain observations accurately for various reason like :

a. Inability to follow sharp colour changes

b. Inability to fix the exact end point colour and sometimes overshooting the end point

c. Persistent use of an off- centre position in observing the swings of the balance

d. Wrong reading of the meniscus in burettes, pipettes etc., in the volumetric analysis particularly in the case of coloured liquids

e. Lack of knowledge about the instrumental errors that are likely to occur due to overheating, ageing of the instrument, change in the order of addition of the reagents, reagent errors, type and quality of reagents to be used for each purpose, use of wrong glassware for analysis as in the case of boron estimation or phosphorus estimation, use of standard curves prepared much earlier etc.

iii. *Errors in methods* : This is one of the important sources of error. Usually these errors are due to the incorrect method of sampling, wrong choice of a reaction and incompleteness of the reactions at some stage or other.

The contribution of these are more in the case of biological samples as many side reactions occur in addition to the interference by other elements. In gravimetric analysis these errors arise due to the solubility of the precipitates, coprecipitation, post precipitation, decomposition, volatilisation of the weighed form or due to the improper ignition of the precipitate. In volumetric analysis in addition to the solubility of the precipitate in the wash solution, errors may also arise due to the failure of reactions to proceed to completion, occurrence or inducing of side reactions in biological systems when the element has to be precipitated first and then washed and titrated as in the case of phosphorus, calcium or potassium estimations of biological materials. Also simultaneous reactions with other elements and radicals present in the solution that is being titrated may also occur. The analyst has to be aware of these possibilities if the errors have to be minimised.

iv. *Additive and proportionate errors :* The value of the additive errors is independent of the amount of the constituent present as loss of weight in the crucible during ignition, errors in weighing etc. Absolute values of proportional error depends on the amount of the constituent and may arise due to the impurity in standard substance which in turn modifies the normality as in the case of sodium sulphate coprecipitate with barium sulphate.

B. INDETERMINATE OR RANDOM ERRORS OR ACCIDENTAL ERRORS

The source of the indeterminate errors cannot be determined as they are irregular and also occur at random. Further the errors are very small in magnitude and may be due to various reasons like slight differences in the calibration of the apparatus, variations in room temperature, variations in the temperature of the electrical equipment during the operations, effect of use of the equipment continuously over a length of time for estimating large number of samples, fluctuations in voltage etc. These are random in nature and cannot be quantified definitely. Even if the measurements are made by the same observer with the maximum care under identical conditions, still the values will vary slightly due to these random errors. If large number of analysis are carried out on the same sample, it will be observed that the errors follow a bell-shaped normal curve pattern and in that the small errors occur more frequently than large errors. Positive and negative errors occur in equal magnitude. The elimination of the contribution of these errors is not practicable.

However, the contribution of these errors to the final volume can be minimised by determining the possible sources of the error and eliminating them if possible or by making allowance for it by adopting some of the following techniques;

i. *Calibration of apparatus, weights and equipments* : One of the important sources of error is due to errors in calibration of glassware, weights etc. Normally exacting standards are insisted in calibrating the certified apparatus and also for issuing the certificate and that is one of the reasons for very high cost of these apparatus. The certificates are classified as "A class", "B class" certificates and they furnish the maximum limits of errors that are likely to occur. When one considers the other random errors, the contribution due to these in the case of certified glassware can be completely ignored. But these certified apparatus are costly, cannot be purchased in bulk for use in big laboratories and that too for routine quality control work where hundreds of analysis have to be carried out everyday and the breakages, discarding of apparatus due to various causes etc., are too many and hence costly for routine work. Similar is the case with college laboratories. In such cases, a set of certified glasswares may be purchased and used to standardise others. Depending upon the accuracy expected in each, separate ones are used or allowances made for the errors due to calibration.

Similarly in the case of weights also, for the same reason ISI certified weights or gold weights cannot be issued for regular work in large laboratories. Hence the weights in other weight boxes are standardised as soon as they are purchased by comparing them with the standard weight. One of the common mistakes done in this regard is that the analyst assumes that no change in the weights occur over time. Unfortunately the acid and alkali fumes in the atmosphere of the laboratories corrode the metals and significant changes have been noticed even within six months. Hence the weights have to be checked with the standard weights periodically preferably once in two months.

Use of filter papers

For each estimation, particularly in gravimetric analysis different types of filter papers are recommended based on the nature of precipitate formed, wash liquids, rapidity of filtering needed, whether the filter paper is to be ashed later or added to the flask during or near the end of titration etc. Use of other quality filter papers invariably results in introducing errors due to creeping of precipitates, passing of precipitates through the pores, also longer time of contact with the wash liquid due to slow movement and hence greater solubility of the precipitate in the wash solutions, and time for side reactions to occur etc. Hence in all these estimations, only the recommended class of filter paper should be used and this will minimise the errors which cannot be quantified.

Quality of reagents

This is the most important factor an analyst has to consider if the errors have to be minimised. The weight of the substance needed for preparing the standard solutions is calculated based on the equivalent weights and on the assumption that the purity of the chemical is 100%. The presence of impurities due to use of a poor quality chemical for the preparation of the solution affects the final normality. In the case of secondary standards, the error is reduced or eliminated because they are standardised against a primary standard before use. But if impurities are present in the primary standard itself, then they produce errors at every step as all subsequent estimations are based on the normality of the primary standard. Hence in all such cases, reagents of highest purity should be used.

Also in case of analysis of microquantities of elements like copper, zinc, manganese in soils, plants and water, the reagents used should be free of that element and preferably double distilled. Even the distilled water has to be checked as they are collected from stills made up of tubes of the same metal or element which are likely to contaminate the waters. For these analysis glass distilled water has to be used. Also it is preferable to redistill the acids and use only redistilled acids for these analysis. Similarly in boron estimations, boron free glassware alone should be used. For phosphorus estimations, fresh glasswares should not be used without adequate cleaning as soluble silica will contribute to the development of colour in the same way as phosphorus. Lack of knowledge about various classifications of quality of reagents has led to considerable amount of wrong information being published in the scientific literature. This can be avoided by using the proper reagents as well as glasswares. For chromotagraphic/ spectrophotometric estimations the reagents used should be of still greater purity than those used for volumetric analysis as ultramicro - or micro quantities are estimated in these estimations and any error committed get magnified by large number of times depending upon the dilutions made.

Running of blank determinations

One method of overcoming the effects of indeterminate errors is to carry out a separate determination with all the reagents as required in the procedure for the test solution. Instead of the test solution, an equal quantity of distilled water or the same solvent used for making up the extract may be added and analysed. The values obtained may be deducted in case of volumetric or gravimetric analysis. In case of instrumental analysis, the blank may be used to adjust the zero or 100 in the instrument before introducing the unknown.

Running a control determination

A control can be run under the same conditions and using the formula given below the correct value of the unknown may be calculated :

$$\frac{\text{Value found for the standard}}{\text{Value of the unknown}} = \frac{\text{Weight of the constituent in the standard}}{X}$$

Use of check samples

To detect and overcome errors due to fatigue of the equipment, voltage fluctuations, leakage in any part of the instrument, already analysed samples can be introduced either randomly or in a regular manner and then the predetermined and currently estimated values compared or cross - checked. If any significant difference is noticed, then the source of error and problem should be traced and rectified before proceeding with other samples.

Use of independent method of analysis

The accuracy of the technique and result is cross verified by carrying out the analysis by another standard method of estimation as in the case of iron which can be done by both gravimetric and volumetric analysis. This procedure can be adopted for test verification of a few samples and if there is a regular pattern in the differences, the reason for which are not determinable by any known method, then a correction factor may be added in the calculations for rectifying the contribution of the indeterminate errors.

Use of parallel estimations

Carrying out the analysis in duplicate or triplicate can help to check and identify the introduction of new sources of errors but a good agreement between the duplicates or triplicate values should not lead the analyst to the feeling that the values are correct as they are being carried out simultaneously and naturally the same indeterminate errors may be present in all the estimations. Hence these estimations are carried out only as a check on the accuracy or analyst's capacity and also to spot out any sudden variations.

Standard additions

In biological estimations, correct estimation of the exact quantity of the constituents present is difficult due to the inability of the extractants to remove all of the material either because the constituent is bound to the substance or substrate or has got occluded or the extractant is not able to penetrate through the whole mass and extract it or the rate of extraction is very slow. In such cases the amount of indeterminate errors will be more. Hence a known amount of the constituent in pure form is added to the sample and after sometime, the sample is analysed for the total amount of the constituent extracted by the extractant. From this the percentage recovery by the particular method is estimated and using this, a factor is arrived and that factor is used in the calculations for correcting the contribution of the errors and arriving at the most probable value. Though this technique of "SPIKING" the sample gives fairly reliable estimation in pure samples, the validity of this in many biological estimations is still questionable. At the same time this is the closest approach to the most probable value that one can attempt in view of the limitations of the available techniques.

Isotopic dilution

This is a very useful tool to estimate and overcome the indeterminate errors' contribution but this has the limitation that many organisations do not have facilities and also in number of cases suitable radioactive isotopes of the required characteristic are not available. Wherever it is possible, it could be tried. In this a known weight of the element being determined containing a radioisotope/ stable isotope is mixed with the sample and the element is isolated in pure form. Radioactivity of the isolated compound or the stable isotope is measured and compared with that of the added element and from this the weight of the element is calculated.

CHAPTER - 5

INDICATORS

In volumetric analysis, the quantity of the unknown constituent is determined by reacting a known quantity of a standard solution with the unknown solution. The point at which equivalent quantities of the reacting materials have been mixed is called the EQUIVALENCE POINT or STOICHIOMETRIC END POINT. Theoretically at this stage, the quantity of the titrant added is stoichiometrically equivalent to the substance under test. Experimentally an observable change in the property is chosen so that the change in that property coincides as nearly as possible with the theoretical equivalent point. Properties like sudden change in colour, pH, conductivity, turbidity have been chosen to indicate the equivalence point. Rarely the end point is found to coincide perfectly with the theoretical equivalence point.

In majority of the titrations, foreign substances are added to the solutions to indicate the change in this property. These substances which are added, are called "INDICATORS". Only in a few cases, the titrating solution itself indicates the end point by a change in its colour. Potassium permanganate is one such solution. This reagent when used in acid solution as a titrimetric solvent, is reduced to almost colourless manganese. When the redox reactions is complete, the next added drop or fractional drop of the permanganate imparts a pink colour to the solution. The titrating solution itself may be used to indicate the end point if it has a particular colour that is not masked by the presence of other constituents. In these cases, the titrating solutions are used up during the titrations and the colour disappears till such time the titration reaction is not complete. Once the reaction is completed, the first trace of excess reagent imparts a colour which can be observed visually. Rarely I_3 and $Cu(NH_3)_4^{2+}$ have been employed as indicators just as permanganate, though they could be used.

Among the substances added, litmus is one which is qualitatively red in acidic solution and blue in alkaline solution. Actually the colour change occurs in the pH range of 4.3 to 8.3. Other substances that are added to give a colour change are very highly coloured dyes that undergo a striking change in colour when going from a solution in which one reactant is in excess to one where the other is in excess. The change could be from one distinct colour to another colour (for example orange to yellow) or from colourless to distinct colour (colourless to pink) or vice versa.

Majority of indicators are two colour indicators because their acidic and basic forms are coloured. The acid - base indicators are themselves weak acids or weak bases and are weaker than those titrated or used as titrants. Hence the indicators do not react permanently with the titrant until the main reaction is completed. The indicators must be selected in such a manner that they correspond closely to the end point of the pH of the neutralisation reaction. The molecular and ionic forms of these indicators have different colours.

Formation of a soluble product of distinct colours : Anions which combine with the first excess of titrating reagent to produce a coloured precipitate or complex ion is also often employed as an indicator. For instance in the Volhard method for silver, the precipitation titration of silver ions with thiocyanates (CNS^-) requires the use of ferric ion (Fe^{3+}) as an indicator. When the precipitation of silver thiocyanate (AgCNS) (white) is complete, the ferric ion unites with the first excess of the thiocyanate to form a Fe(III) - SCN - complex which has deep red colour.

$$Fe^{3+} + CNS^- \rightarrow Fe(CNS)_2 + \text{(red colour)}$$

The acid base indicator acts as a pH indicator since it exists in two forms. The acid forms possess one colour and by a loss of proton is converted to a different coloured base form.

$$\underset{\text{colour A}}{\text{In A}} \rightarrow \underset{\text{colour B}}{\text{H}^+ \text{ In B}}$$

In A is the acid form with colour A and In B is the base form of the indicator with a different colour B. Methyl red is a typical acid - base indicator. Red coloured forms predominate at pH 4.2 and below whereas the yellow base form predominates at and above pH 6.2

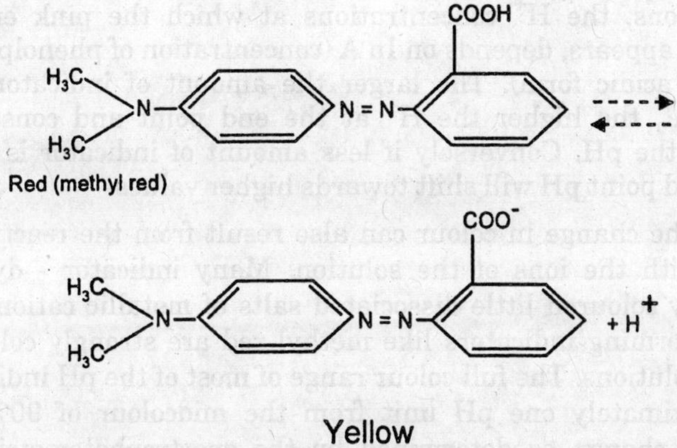

The behaviour of one colour indicator is different and more complicated. Phenolpthalein is one such indicator. It is colourless in acidic form and pink in basic form and both are in equilibrium with each other as shown below:

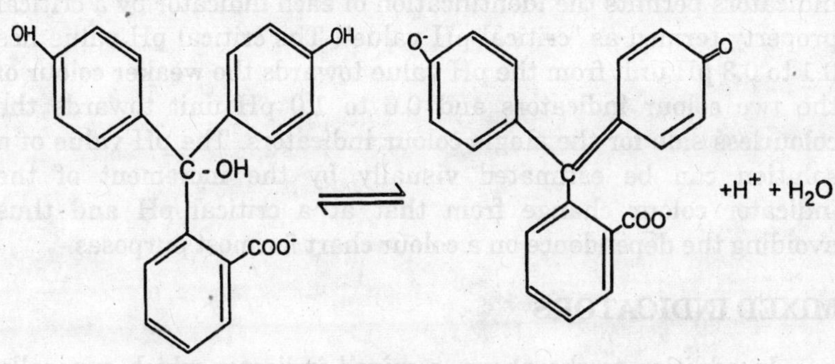

Phenolpthalein is employed as an indicator for the titration of strong acids with strong bases. The end point is indicated by the first permanent appearance of a faint pink colour throughout the solution. The intensity of pink colour which is detectable by the eye is very small and quite reproducible regardless of the total concentration of acids and bases involved in the titration. The minimum detectable concentration of the pink form can be found out from the equation:

$$K\,In = \frac{(H^+)\,(In\,p)}{(In\,A)} \quad \text{or} \quad (H^+) = (K\,In)\,\frac{In\,A}{In\,p}$$

where K In is the indicator constant. As could be seen from the equations, the H^+ concentrations at which the pink end point colour appears, depends on In A (concentration of phenolphthalein in its acidic form). The larger the amount of indicator that is present, the higher the H^+ at the end point and consequently lower the pH. Conversely if less amount of indicator is present, the end point pH will shift towards higher values.

The change in colour can also result from the reactions of a dye with the ions of the solution. Many indicator - dyes form weakly coloured little dissociated salts of metallic cations. Weak base forming indicators like methyl red are strongly coloured in acid solutions. The full colour range of most of the pH indicators is approximately one pH unit from the midcolour of 90% of the colour change as determined by the spectrophotometric colour analysis of the indicator curves and this is due to the relationship of colour to the dissociation constant of the indicator materials. A tenfold increase in H^+ concentration (a decrease in 1 pH unit) changes the concentration ratio of salt/acid by 10/100 or by 90% towards a complete change to acid colour. The behaviour of pH indicators permits the identification of each indicator by a critical property termed as "critical pH value". The critical pH value lies 0.1 to 0.3 pH unit from the pH value towards the weaker colour of the two colour indicators and 0.6 to 1.0 pH unit towards the colourless side for the single colour indicators. The pH value of a solution can be estimated visually by the increment of the indicator colour change from that at a critical pH and thus avoiding the dependence on a colour chart for most purposes.

MIXED INDICATORS

In addition to the above, a mixed indicator which generally consists of a conventional acid-base indicator with an added organic dye is sometimes used to increase the sharpness of the visible end point. The added dye may be another indicator but more commonly it is a substance not affected by pH. A proper substance of the latter blocks out or masks certain wavelengths of light common to both colours of the acid-base indicator. Essentially only the components of Colour A and of Colour B which differ from each other are observed by the analysts. Consider an indicator whose colour A is blue green and colour B is yellow green and has an added dye which absorbs the green light. The mixed indicator would exhibit a colour change from blue to yellow green obtained without the added dye.

Mixed indicators may involve even more complex colour relationships. Methyl orange and xylene cyanate FP comprise a common mixed indicator. Methyl orange appears yellow in alkaline medium, xylene cyanate absorbs yellow and orange light very strongly so this mixed indicator appears green in alkaline solution. In an acid medium, methyl orange absorbs light in the green region, xylene cyanate still absorbs the yellow orange region and the colour of the mixed indicator is blue red. In the intermediate region which is at approximately pH 4, the solution is nearly colourless because of the transmission of complementary colours.

Another mixed indicator, methyl purple which is commonly used, consists of methyl red (red below pH 9.3 and yellow above 6.2) mixed with an inert blue dye stuff. This indicator gives a very pronounced colour change and this change occurs over a very narrow pH range. The blue dye with the red of the indicator gives a purple colour below pH 4.8 and with the yellow of the indicator a green colour above 5.4. Between pH 4.8 and 5.4, the red - orange of methyl red and blue are complementary giving a light grey colour over a short range of pH. Because of this narrow range, extra care has to be taken to see that the grey colour is not allowed to pass unnoticed.

The indicator sensitive to changes in pH will change colour when it changes from acid to base form. If the indicator and the background colours are complementary at a certain concentration ratio of the two forms of indicator, then it will be easier to observe the change in colour over a narrow pH range. The complementary colours give a light grey or nearly colourless appearance over only a few tenths of a pH range.

Sometimes a dye insensitive to pH changes is mixed with an indicator to give a background colour that is complementary to the indicator at a certain pH or else the background accounts for a certain colour of the indicator at a certain pH. For example when phenolphthalein which is colourless in acid and red in alkaline form is mixed with the methylene grey which is greenish blue in any pH, there is a colour change from green in acid to pale blue at pH 8.8 and violet at pH 9.0. The following are a few of the mixed indicators used in volumetric analysis :

MIXED INDICATORS

Mixed indicator				Acid	Base	pH change
Methyl yellow	+	Methylene blue	1 : 1 alcohol	Blue	Green	3.2
Methyl orange	+	Xylene cyanol F.B.	2 : 3 alcohol	Red	Green	3.8
Methyl orange	+	Indigo caramine	2 : 5 water	Violet	Green	4.1
Methyl orange	+	Bromocresol green	1 : 5 water	Orange	Blue-green	4.3
Methyl red	+	Bromocresol green	2 : 3 alcohol	Red	Green	5.1
Bromocresol green	+	Chlorophenol red	1 : 1 water	Yellow green	Blue violet	6.1
Methyl red	+	Methylene blue	1 : 1 alcohol	Violet	Green	7.0
Phenol red	+	Bromothymol blue	1 : 1 water	Yellow	Violet	7.5
Cresol red	+	Thymol blue	1 : 3 water	Yellow	Violet	8.3
Thymol	+	Phenolphthalein	1 : 3 alcohol (50%)	Yellow	Violet	9.0
Phenolphthalein	+	Nile blue	1 : 2 alcohol	Blue	Red	10.0

*Total concentration of indicator is about 0.1% in each case

UNIVERSAL INDICATOR

Universal indicator is another type of mixed indicator not generally used for titration but only for approximate pH measurements over a wide range. The table below gives the composition of some of the universal indicators. The first of these show colours in the order of the spectrum, starting with red at pH 3 and passing through orange at pH 5, yellow at pH 6, green at pH 7.5, blue at pH 9 to violet at pH 10. Accurate pH measurements may be made with such mixtures if they are calibrated with known buffer solutions. Test papers impregnated with universal indicators are available for rough pH testing and they are extremely useful.

COMPOSITION OF CERTAIN UNIVERSAL INDICATORS

Constituents	Indicator Numbers		
	1	2	3
Methyl yellow	–	0.02	0.010
Methyl orange	0.030	–	–
Methyl red	0.015	0.30	0.013
Bromothymol blue	0.030	0.40	0.160
Naptholphthalein	–	0.40	0.160
Cresolpthalein	–	0.40	0.240
Phenolpthalein	0.150	0.20	0.400
Alizarin yellow G	–	–	0.300

*Preparations given are in grams/liter of 60% alcohol

ADSORPTION INDICATORS

These indicators are used in certain precipitation titrations. Depending upon the first ion of titration and hence upon the ion of the precipitate that is primarily adsorbed before and after the stoichiometric point, adsorption or desorption of the indicator, counter ions occur immediately upon reversal of the charge of the adsorbed ions of the precipitate.

The adsorption or desorption process is accompanied by a change in the structure of the indicator resulting in a change of colour. To keep the precipitate finely dispersed in the solution, protective colloids are often added before starting the titration. This method has been used in the titrimetric precipitation of silver halide using fluorscein and related compounds in Fajan's method of chloride. Organic acids such fluorscein or organic bases such as rhodamines produce ions in solution of one colour or nearly no colour. When the ions are adsorbed on a precipitate of opposite charge, the ion of the acid or base change colour. If chloride ions are being titrated with silver ions, the precipitate first formed will adsorb chloride ions and the silver chloride particles will acquire a negative charge. Such particles will not adsorb the fluorscein ion which is also negatively charged. At the equivalent point, the chloride and silver ions are adsorbed equally and a very slight excess of silver ion changes the charge on the precipitated particles to positive ones. The positively charged particles attract and adsorb the negative fluorscein ions and the latter becomes pink in colour indicating that the end point has been passed. Similarly positively charged ions can be adsorbed by negatively charged particles, the change in colour on being adsorbed indicating that an equivalence point has been passed.

FLUORESCENT INDICATORS

One of the limitations of the conventional acid - base indicators is that they cannot be used in strongly coloured or highly turbid solutions. This limitation can be overcome by using an indicator or weak acid or base which fluoresces in the acid form but not in the basic form or vice versa. (eg) X - napthalamine. The free amine shows a blue fluorescence in ultraviolet light whereas the salts of this do not. X - napthylamine fluoresces above pH 4 but shows little or no fluorescence below pH 4. Theoretically the intensity of fluorescence could be used to measure the pH within a certain range but due to other reasons, fluorescent indicators are used only in certain titrations.

FLUORESCENT INDICATORS

Indicator	pH range	Acid Fluorescence	Base Fluorescence
Eosin	0.0 – 3.0	None	Green
4-Ethoxyacridane	1.2 – 3.3	Green	Blue
Salicylic acid	2.0 – 4.0	None	Blue (violet)
B- Naphthylamine	3.4 – 4.4	None	Violet
X- Naphthylamine	3.4 – 4.8	None	Blue
Dichlorofluoresceine	4.0 – 6.6	None	Green
3,6, Dihydroxyxanthane	5.4 – 7.6	None	Blue
Umbelliferone	6.5 – 7.6	None	Blue
Coumaric acid	7.2 – 9.0	None	Green
B- Napthol	8.0 – 9.0	None	Blue

FACTORS AFFECTING THE DETECTION OF END POINTS

The efficiency and accuracy of detection of end points using an indicator depends on number of factors like direction of titration, concentration of solution, concentration of indicator, temperature and nature of solvents. Their influence on the accuracy of the estimations varies and by proper choice of the conditions, the exact end point can be detected easily.

a. *Direction of titration* : Indicators with pH range nearer 7 can be used without any concern about the direction of titration. In titration of strong acids with a strong base using methyl orange, the end point is on the acid side of the pH range of methyl orange. Titrating the base with an acid gives the end point on the alkaline side of the methyl orange but the difference is not very significant.

In strong acid - weak base titrations where the equivalence points are less than pH 7, the indicators chosen when titrating acid with base should have a pH range from a few tenths of a unit below the equivalence point to a higher pH. Conversely when titrating weak base with a strong acid, the indicator should have a pH range from a few tenths of a pH unit above the equivalence point to a lower pH.

In weak acid - strong base reactions, the equivalence points are at a pH greater than 6 and in such cases indicators like phenolphthalein should be preferred. The pH range for acids with $K_a > 10^{-5}$ is 7 to 12. For weaker acids with $K_a > 10^{-6}$ the range is reduced to 8 to 10. The pH range of 8 to 10.5 will cover most of the titrations and this permits the use of thymol blue, thymolphthalein or phenolphthalein. During the titration the indicator should show a visible change of colour when 1/10 of it has changed from one form to another. Therefore for weak acid - strong base titration, the indicator should have a few tenths of a pH unit below the equivalence point to above the equivalence point. Titrating from the alkaline side of acid, the indicator should have a range from a few tenths of a pH unit above the equivalence point to below the equivalence point. Generally the direction of titration makes very little difference in the choice of indicator except in rather dilute solutions (<0.01 N). In case of weak base and strong acid, the pH at the equivalence point is computed from the equation.

$$pH = 0.5\ pK_w + 0.5\ pK_o - 0.5\ pC$$

The pH range for bases with $K_b > 10^{-5}$ is 3.7 and for weaker bases $K_b > 10^{-6}$ is 3.5. For these titrations, the suitable indicators are methyl red, methyl yellow, bromocresol green and bromothymol blue.

In case of weak acid - weak base, there is no sharp rise in the neutralisation curve and hence no simple indicator can be used. Titrations of this type should be avoided if possible. A mixed indicator which exhibits a colour change over a limited pH range may be employed. For example neutral red - methylene blue can be used for ammonia solution and acetic acid. The approximate pH at the equivalence point can be computed from the equation

$$pH = 0.5\ pK_w + 0.5\ pK_a - 0.5\ pK_b$$

As a rule whenever an indicator does not give a sharp end point it is better to prepare an equal volume of comparison solution containing the same quantity of indicator, the final products and other components of the titration as in the solution under test and to titrate to the colour shade thus obtained.

b. Concentration of solutions : In any titration, there should be a change of approximately two units of pH at or near the stoichiometric point due to the addition of a small volume of the reagent. For example at pH 4.2 and below, methyl red is present in red acid form and at pH >6.2 as a yellow base form. The range of pH values over which the acid base indicator change is usually called the pH TRANSITION INTERVAL or pH TRANSITION RANGE. This range is often 2 pH units although it may be smaller for a few indicators. The reason for this is that the human eye responds more readily to some colours than to others. At the same concentration, certain indicators are naturally more intensely coloured than others. If the total concentration of two coloured indicators is increased, then the individual concentration of both the acidic and basic forms increase simultaneously and hence the pH transition zone remains unchanged.

For example in a strong acid - strong base titration involving solutions of strength stronger than 0.1 M, any indicator having a range between 4.5 to 9.5 may be used, that is any indicator between methyl orange (3.2 to 4.4) and phenolphthalein (8.3 to 10.0). As dissolved carbon dioxide acts as an acid on phenolphthalein, carbon dioxide should be removed by repeated boiling (3 to 4 times) while still acid and the solution titrated when cold. If this is not possible an indicator with a range < 5 pH should be employed. With solutions of 0.01 M or lower strength the pH range is somewhat smaller (5.5 to 8.5) as could be seen from the titration curve using different strengths.

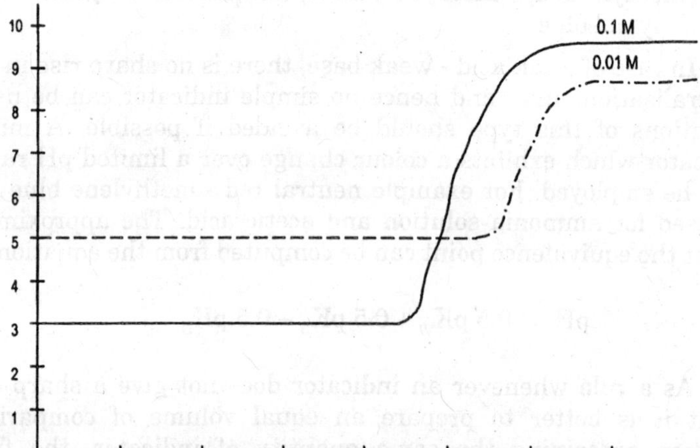

Volume of NaOH used

c. Concentration of indicator : The change of colour of an indicator from one to another actually uses up or liberates some hydrogen ions. The quantity of hydrogen ions that is consumed or produced must be negligible in comparison to the amount of acid or base involved in the main titration reaction. Otherwise a significant error is introduced. The magnitude of this INDICATOR ERROR increases as the contribution of the indicator is increased. Therefore the indicator concentration should not be larger than that required to render the end point colour change visible. The error is often slight but of measurable magnitude and this is one of the reasons why experimental end points and critical equivalent points do not coincide perfectly.

A correction for indicator error can be made by means of a blank titration. A solution of the same composition of the real sample containing the indicator but not the acid or base being determined, is titrated to the same end point as in the real titration and the small volume of titrant is subsequently subtracted from the total volume used in the real titration. If the error is appreciable, a blank should be run with the proper quantity of the indicator to be used.

d. Temperature : Temperature affects the value of K_w and K_{ind} for indicator. The indicator should be calibrated if it is to be used at temperatures other than the room temperature. In most cases, the ionisation constant first rises with rising temperature, then passes through a maximum and falls again. For carboxylic acid, the median is not far from room temperature so that in general these ionisation constants do not change much with temperature in the usual working range.

The main factor causing the fall in ionisation with raising temperature is the effect of increasing temperature on the dielectric constant of water which is to decrease it.

$$NH_4^+ + H_2O \rightarrow NH_3 + H_3O$$

The azo indicators such as methyl orange and methyl yellow show considerable increase in dissociation as the temperature rises. A solution containing methyl orange which had been adjusted to give the orange-red transition colour of the latter turns yellow on heating. The pH range of visible colour change decreases from 3.1 to 4.4 at 18▯C to 2.5 to 3.7 at 100▯C. Now in the dissociation of the azo indicators, a proton is separated from the tertiary ammonium ion leaving an unchanged residue just as in the dissociation of ammonium ion.

In the case of phenolphthalein, for titrating carbonic acid or carbonic ions to bicarbonate, the amount of base or acid taken to reach the end point is not affected by temperature. Evidently the effect of temperature on the ionisation of carbonic acid and on the ionisation constant of the indicator just compensated one another.

e. Effect of solvents: Solvents other than water may cause alteration in the value of K_{ind}. If nonaqueous solvents are involved, the indicator should be calibrated for the conditions of analysis.

OTHER METHODS FOR DETECTION OF END POINTS

In addition to a change in colour, certain properties like conductivity, single electrode potential, fluorescence effects due to the radiations of light of one wavelength on particles suspended in a light when light of different wavelengths impinges upon them have been used to detect the end points.

a. Electrical methods: The end point of titrations can be found on the basis of changes in the electrical properties of the titration mixtures. Here a graph of the electrical property is made against the amount of titrant used. The stoichiometric point is indicated by a rapid rate of change of the electrical property measured.

Instead of coloured indicators, pH meter could also be used for detecting the end points when a gradual change in pH occurs at the end point i.e. when titrating very dilute solutions or weak acid and bases or when the pH of the end point is unknown and a choice of indicator is difficult to make. The pH meter is a modified potentiometer designed to read the pH of solution directly when the standard electrodes are dipped into the solution. The pH is plotted in a semilog paper against the volume of the reagent added. The equivalence point is the point on the volume axis when the rate of change of pH is at a maximum. The sharpest point in the curve is known as the inflection point and this gives the end point.

The end point may also be detected by sharp change in the electromotive force of a suitable electrode near the stoichiometric point due to the addition of small increments of the titrant.

By measurement of the conductance and following the rate of change of conductance, the end point can be detected. Removal of ions from the solution by neutralisation, precipitation or completion of reaction results in marked 'Breaks' in the rate of change of conductance with the amount of the titrant added near the end point. In the chloride - silver ion titrations, the conductance of the solution reaches a minimum at the equivalence point. Similarly the rate of change of flow of current in a cell during titration can also be used for finding out the equivalence point.

b. Type clear point method: The end point can be detected by observing the point at which all the colloidal particles flocculate and the solution above the precipitate becomes clear and free of all colloidal particles. As precipitate forms, they adsorb from the solution whatever ion is common to the precipitate more strongly than they adsorb any other ions. Only one ion common to the precipitate is in excess in the solution at the beginning of the titration. The ion is adsorbed on the small particles of the precipitate, causing them to become charged and to remain suspended. The equilibrium between ions in the solution and ions adsorbed on the precipitate particles is disturbed by the titration. The concentration of the ions in solutions gives the charge to the precipitated particles and is reduced as the equivalence point is reached. At the equivalence point, ions of opposite charge are in equal amounts. Therefore the two oppositely charged ions are adsorbed at the same rate leaving the colloidal particles unchanged. The colloidal particles unite with each other to form larger particles which settles leaving the solution clear. It takes little time for the adsorbed ions to diffuse to outside of the floccules formed. It is therefore easy to pass the end point.

c. Turbidity methods : In certain cases, occurrence or disappearance of turbidity may be made use of in determination of the end point. Some changes in the turbidity may indicate the end point. Certain weak acids and bases are highly insoluble while their salts are soluble. Such substances can be used as indicators for a precipitate will appear or disappear as the pH is changed or in the appropriate range. The practical value of such indicators lie in the fact that the turbidity can be made to appear very sharply in a very narrow pH range. For example in the case of m-toluene azonitrosoacetyl-toluene the change is in the pH range of 9.3 (turbid) to 9.4 (clear).

The appearance of the precipitate may indicate the end point in a titration such as that of a cyanide ion titrated with silver ions. Excess cyanide forms a soluble complex ion with the silver until the excess is gone. Then a precipitate forms.

$$2\,CN^- + Ag^+ \rightarrow Ag(CN)_2^- \text{ (soluble)}$$

$$Ag(CN)_2^- + Ag^+ \rightarrow 2\,AgCN \text{ (white precipitate)}$$

The end point is the first detectable turbidity or opalescence in the titration mixture.

d. Clear turbidity method : The equivalence point is indicated by removing two small samples of the solution from the titration flask and adding a little of a solution of one ion being precipitated to one sample and a little of solution of the other ion to the other sample. On either side of the equivalence point, one sample will become more turbid than the other. At the equivalence point, both the samples will show an equal amount of turbidity. Carefully standardised test solutions of ions may be added from the burettes. Then this gives precise results.

OTHER USES OF COLOUR INDICATORS

a. Measurement of pH : As commercial glass electrode pH meters are available, indicators are rarely used now for pH measurements. However as these indicators are cheap and easy to use, they are still preferred by many. Indicators can give relatively rough pH estimations like universal indicator papers which are moistened with solution to be tested and its colour compared with a chart. Greater accuracy can be obtained by adding a measured amount of the indicator solution to the test solution, mixing and comparing the colour visually with the solutions of the indicator in the standard buffer solutions. Colours could be matched with the help of a comparator.

In a comparator the coloured glass discs are mounted on a wheel held upto the light. Then the wheel is turned until the colour of the glass in line of sight matches with that of the solution under the test kept in a test tube or in a rectangular cell. The eye piece contains a prism which brings the two beams of light, one from the solution and the other beam from the coloured glass together and helps to compare them easily. In cases of coloured solutions, the comparator has two more cells, one containing some of the original solution placed behind the coloured glass and the other containing water placed behind the test solution. Sets of glasses are available for several different indicators and are generally made to stimulate colour at intervals of 0.2 pH. By interpolation, the readings close to 0.1 pH can be obtained. A comparator of this type can be used for all kinds of colorimetric determinations where a quick routine method of medium accuracy alone is needed.

CHAPTER 6

COMMON APPARATUS AND EQUIPMENTS IN AN ANALYTICAL LABORATORY

The various methods of analysis described in this book for waters, fertilizers, manures, plant materials and soils could be carried out in any laboratory as the apparatus and other equipments needed for the analysis are simple and commonly found in any analytical laboratory. Further all these are manufactured locally in any country and also not very costly. However their utility, accuracy and working life depends on the care bestowed by the analyst who is using them and this indirectly reflects on the investment to be made by the organisation in maintaining and running the laboratory and also on the cost per analysis. In view of the present emphasis on the use of low cost, high productive technology, the use and maintainence of the important apparatus and equipments are discussed here as this calls for practically no investments but at the same time highly remunerative to the organisation in terms of time, money and its standing among the other similar institutions. The information provided here could even be used for establishing a new analytical laboratory and in case of the established laboratories, to understand the potentialities and uses and thus increase their efficiency. The following are the common apparatus and equipments used in a laboratory.

a. Bunsen burner: It is one of the essential items for any chemical laboratory. It consists of a base with a side tube ending in a jet over which is screwed a metal tube. A short metal ring is slipped on it and by rotating it the air supply to the inside of the tube can be regulated. The side tube can be connected to the main gas supply. The stream of coal gas passing out of the jet reduces the pressure and when the air hole is open, air is sucked in through it into the burner tube to get mixed with the gas. The mixture burns when ignited at the top of the tube of the burner giving a non- luminous flame.

As the outer zone of complete combustion increases in thickness, the luminous region gradually disappears and the flame becomes shorter and steady, as inner cone of pale blue colour is seen. The non-luminous flame has two cones - a pale inner cone of gas with the incipient combustion (reducing zone) which turns green and shining when a large volume of air is admitted. The second cone is a still paler blue outer cone of complete combustion (oxidising zone) which is constant in size. The tip of this outer zone is hottest part of the flame. The more the air is sucked, the hotter is the flame but however this cannot be extended much because when the ratio of air to gas is wide, the inner cone becomes green and the flame begins to "roar". If still more air is admitted, the flame "strikes back" from the mouth of the burner through the tube to the mouth of the jet. Here the acetylene is formed and the burner will become very hot. Hence a judicious proportion of gas to air has to be maintained. In case of incomplete combustion, the flame becomes luminous because of the presence of carbon particles in the flame.

Nowadays in view of high cost of producing a coal gas, problems connected with the maintenance of the pipelines, fluctuations in pressure due to blocking in pipelines, fluctuations in heat of the fire hearth and changes in the quality of fuel fed into the fire hearth, dependency on electricity supply, or fire hearth and other problems in production of coal gas, many laboratories are switching over to the use of petroleum cooking gas like Indane, Bharat, Burshane gas etc. The gas is supplied in different sized cylinders to suit the needs of the customers. As it is supplied in cylinders under high pressure, these cylinders have to be connected to the burners through appropriate regulators. Also the ordinary bunsen burners have to be slightly modified to give a non-luminous flame as the gas that is stored in the cylinders is butane.

b. Balances: The basic instrument needed by an analyst for carrying out either volumetric or gravimetric analysis is a balance for weighing. It works on the equal arm principle and has a beam and two pans attached to the two ends of the beam with the fulcrum and the pointer located at the center of the beam. Various improvements are constantly being made to increase the sensitivity of the balance as well as the speed of measurements. In all the cases, the object to be weighed is kept on one pan and compared with the weights kept (or added) on the other pan. The following are few of the different types of balances commonly found in the laboratory :

i. *Physical balance*: This is used to weigh objects upto a gram or slightly less weight. It is the least sensitive of all the balances used in the laboratory. When the weight of the material needed is only to the gram weight level or when the weight of the test materials exceeds 100 gm, then this could be used as the errors due to the neglect of the fractional weights will be less than one per cent.

ii. *Chemical balance* : Among the balances in a chemical laboratory, chemical balance is the one most commonly used. It weighs samples to a milligram level when used in conjunction with a rider.

iii. *Single pan balance:* This is much more sophisticated balance and can weigh upto a fraction of milligram and that too with greater speed than a chemical balance. In this the weights are removed or added by turning a knob. The optical device and the digital mechanism also help to increase the speed of measurements. In view of the greater sensitivity and other mechanisms, the single pan balance is costly and needs greater care and good maintenance for assuring accuracy and longer working life. But the amount invested in acquiring the balance is recovered in a short period if the laboratory has sufficient work load daily as the cost per analysis will be considerably reduced when large number of samples are analysed.

iv. *Chainomatic balance :* In this model, the smaller weights, less than a gram, are added through the movement of a chain attached to the end of the beam. The chain is previously calibrated and the raising or lowering of the chain either removes or adds weights and the weight added is indicated by the numbers on the knob. The speed of measurement is also much more than the chemical balance but the chains need to be cleaned and calibrated frequently as they are affected by the fumes in a laboratory.

v. *Key board balance*: This also works on the same principle as a single pan balance except that the weights are removed or added by the touch of the keys at the bottom of the balance, just as in a typewriter. This model also gives greater speed of measurement and precision.

vi. *Damped balance*: This has a device to slow down the swings or bring it to the zero point quickly.

vii. *Microbalance*: This is essentially the same as any chemical balance except that it can be used for weighing microquantities of materials and to smaller fractions of a milligram.

In the choice of balances, two important terms are used to indicate their utility. They are (i) Sensitivity - which indicates the smallest weight that can be measured by the balance. ii. Sensibility - is the number of scale divisions of the zero point or rest point that is changed when one milligram is added to one pan of the balance.

Environment significantly influences the accuracy of measurements. Even a slight drift of air can swing the pointer away and change the resting point of the pointer. Hence it is absolutely necessary that the balances are kept in locations wherein wind, either natural or artificial by means of fans do not enter the room. The rooms where balances are kept should never be fitted with electric fans. As far as possible the balance room should have sufficient natural lighting than artificial lighting by tubelights etc., as the latter when in operation for a long time increases the temperature of the room and thus affects the sensibility of the equipment. Many of the materials weighed in a balance have a tendency to absorb moisture and in a few cases may even change to liquid state if they are hygroscopic materials. Hence the inside of the balance should be maintained in a dry state by keeping desiccants like calcium chloride, phosphorus pentoxide, calcium sulphate, magnesium perchlorate, sulphuric acid inside the balances. Wherever drying agents are kept, they should be checked periodically and replaced whenever signs of their being moist is observed. In case of liquid desiccants like sulphuric acid, care should be taken to see that they do not overflow their containers due to absorption of moisture. Whenever these containers become 3/4th full, they should be emptied and small amount of fresh acid kept again.

Change in temperature inside and outside the balance affects the sensitivity of the balance. Hence use of burners, heating elements or other hot materials near a balance should be avoided. Also only materials at room temperature should be weighed in a balance and on no account hot materials should be weighed or even kept on the pan of the balance. If the moisture free values are needed, then the hot materials may be kept in a desiccator, allowed to cool to room temperature and then quickly kept on the pan of balance and weighed.

The accuracy of measurement by a balance not only depends upon the balance but also on the weights used. The weights are usually kept in a weight box. A common weight box consists of 500, 200, 200, 100, 50, 20, 20, 10, 5, 2, 2 and 1 gram weights and also 500, 200, 200, 100, 50, 20, 20, 10 milligram weights and forceps. The gram weights are usually metal coated alloys while the milligram weights are made of aluminium or any light metal. Normally the weights are expected to be calibrated against standard weights and then only supplied to the laboratories. Still in the interests of analysis and analysts, the weights should be cross checked against either ISI standard weights or gold coated standard weight. The difference between the actual weights and the expected values as per the markings widen with increase in use due to corrosion by acid/alkali fumes present in the laboratory atmosphere, adhesion of colloidal materials, sweat etc., to the weights due to lifting of weights directly by using fingers instead of forceps.

As certified weights are very costly and no laboratory can afford to issue such weights for each balance, it is advisable that each laboratory maintain one or two sets of such weights and periodically cross - check each weight box and discard those which do not conform to the specifications. Also the certified weights should be kept in a safe place where they will not be affected by acid or alkali fumes.

c. Distilled water still : Every analytical laboratory needs large quantities of distilled water for daily use. This is produced with a distilled water still. In this the tap water or water from any other source is allowed to continuously flow through at a constant rate through heating coils. The liquid water is converted into steam and comes at the exit leaving behind the salts. The steam is condensed and collected as a pure water. The amount of distilled water produced varies with the size of the still.

The efficiency and life span of a distilled water still can be improved by taking the following precautions:

i. As far as possible water containing low amounts of salts just like rain water alone should be used for distilled water preparation. If the waters contain high amounts of salts, the coils and other materials inside get coated with salts and the still will get corroded within a short time and will become unfit for further use.

ii. The rate of flow of water should be uniform and continuous and neither too fast nor too slow. Many a time the heating coils burn away due to breakdown in the supply of water caused by the failure of motors or emptying of the overhead tank. Hence extreme care should be bestowed to provide a continuous water supply from an overhead source whenever the water still is working.

iii. Periodically the distilled water collected should be checked for the salt content and also they should be collected and kept in a clean containers after giving two or three rinses with distilled water.

It is a common notion among new comers to the laboratories that for every type of analysis, distilled water alone should be used. This creates considerable amount of pressure on the limited supply of distilled water in a laboratory. This could be overcome by exercising certain amount of discretion in choice of water for use in different analysis. For number of analysis like washing phosphorus precipitate, nitrogen distillation etc., pure rain water collected in covered tanks during rainy seasons and allowed to settle for sometime could be profitably used as they contain very little salts and these salts do not interfere with the analysis. If necessary a blank may be run and necessary corrections made but invariably the amount of corrections to be applied will be very little.

Another method to overcome shortage of distilled water in a laboratory is to use a deioniser or demineraliser. Number of models and with different outflow rates are available now in the market. Further the operation of these equipments is also simple. The water is allowed to pass through two-ion resin beds, one with positive charge and the other with the negative charge and the deionised water is collected at the other end. The positive ions of salts in water replaces the H^+ from the cation exchange resin and the negative ion of the salt of water replaces the OH^- from the anion resin. The replaced H^+ and OH^- combine to yield water which contains practically very little of salts and in majority of instances below the detectable limits. The salt content of the outgoing water can also be tested with the meter fitted on to the demineraliser or deioniser or can be tested with a solu-bridge.

An important point in their maintenance is that once detectable amounts of salts are observed in outgoing waters, the resins have to be regenerated. Each manufacturer supplies the procedure to be followed and it is generally a simple one and a reverse process technique using acids and alkalies, to replace the cations and anions present on the resins. Regeneration of the resins in separate beds is accomplished by passing hydrochloric acid through the cation exchange bed and a basic solution of ammonia through the anion exchange resins. In case of single bed exchangers the two resins are first separated by making use of difference in their density and then regenerated as before. After regeneration they are again mixed as per the instructions of the manufacturer. By observing the instructions of operations, bulk quantities of demineralised water can be produced daily in a short time to meet the daily needs of most analysis.

In certain special types of analysis like micronutrient analysis, ordinary distilled water cannot be used as it invariably contains traces of Cu, Pb, Fe etc. In such cases the distilled water is further distilled in an "all glass still" and the pure water collected in glass containers for specific uses. As the amount produced per unit in these glass stills is very small, the minimum quantity required for each analysis alone should be used.

In cases of emergency and also in teaching laboratories where number of nitrogen distillation sets are available normally, additional amounts of distilled water can be produced by distilling water in these distillation sets. In view of the high cost of these apparatus, they should never be allowed to become dry during distillation and also at the end of each day, the salts formed inside the flask should be cleaned thoroughly with acids even if the same unit is going to be used for the same purpose the very next day. However as the cost of production by this method is very high, it should be resorted to only in emergency circumstances.

d. Glasswares: In a chemical laboratory, different types of glasswares are used. Among them some are very accurately calibrated while the others are approximately quantified. The accurately calibrated glasswares are:

i. Burettes (10, 25, 50, 100 ml) (microburettes 5, 10 ml)

ii. Pipettes (1, 2, 5, 10, 20, 25, 50, 100 ml), micro-pipettes (1 and 5 ml) graduated and nongraduated pipettes; in some laboratories 11 ml and micropipettes of 0.1, 0.2 ml etc., are also used.

iii. Volumetric flasks (25, 50, 100, 250, 500, 1000 and 2000 ml)

iv. Measuring cylinders (1, 5, 10, 25, 50, 100, 250, 1000 ml) - all graduated. Also in some laboratories measuring cylinders of exactly 1000 ml capacity but without spout are also used.

These are expected to be accurately graduated but due to various reasons, many of them do not come upto the required standards. When such defective ones are used, the accuracy of the estimation is reduced though the analyst may be able to get precise concordant values. Hence these glasswares should always be checked against certified standard apparatus before using them. Now a days certified standard apparatus are sold with a certificate from standard recognised institutions indicating the details of accuracy that can be expected with the apparatus that have been supplied to them and the various tests carried out with them. Even these apparatus are classified as conforming to different grades of accuracy like Class A (or I), Class B (or II) depending upon accuracy. Though it would be advisable to use only certified glasswares, in reality it is not practicable as they are four to five times costlier than the ordinary glasswares. Hence it is preferable to purchase one or two certified glassware sets in each category and use them for standardising and grading the available apparatus, depending upon the degree of accuracy. Then these may be used for different purposes depending on the level of accuracy expected.

For achieving greater precision and accuracy, apart from the above major sources of error, other errors due to improper reading of the graduations due to parallax error, presence of air bubbles in the tips of burettes and pipettes while delivering or drawing of solutions, not allowing sufficient time for the liquid adhering to the sides to come down etc. should be avoided. Another important precaution to be observed in the use of graduated glasswares is never to fill up these apparatus with hot solutions and in cases when they have to be necessarily to be filled with hot solutions as in the case of collection of the hydrochloric acid extracts of soils, these solutions after collection should be allowed to cool to room temperature before making up to the volume. Also in certain types of estimations like phosphorus, brand new glasswares should never be used as considerable amount of contamination due to soluble silica will be introduced. In such cases, the apparatus should be washed with acids number of times and then rinsed with distilled water and then only used. Also for certain of the special analysis like boron estimations, only boron - free glasswares should be used and not the ordinary glasswares as they are mostly borosilicates.

ii. Other glasswares : In addition to the graduated glass apparatus, the following ordinary glasswares are also needed for day to day work. Each one of them come in different sizes and also form. There is a wide choice from which one can choose depending upon the method of analysis. Whenever there is a possibility of heating these apparatus over hot flames, it is better to select Pyrex or other heat resistant glasswares that can withstand heating to high temperatures.

The following are some of the glasswares commonly used in a laboratory :

i. Beakers (100, 200, 250, 500, 1000, 2000 ml). Also tall form beakers of 250 and 500 ml capacity.

ii. Boiling tubes and test tubes with rim and without rim.

iii. Conical flasks (100, 250, 500, 1000ml)

iv. Funnels (ordinary and ribbed) (5, 7.5, 10, 12.5, 15 cm diameter)

v. Weighing bottles (round, squat form, tall form)

vi. Watch glass (5, 7.5, 10, 12.5, 15 cm diameter)

vii. Glass scoop

viii. Ice tumbler

ix. Kjeldahl flasks (500, 750, 1000 ml)

x. Glass wash bottles or plastic wash bottles (500, 1000, 2000 ml)

xi. Distillation flasks and distillation sets.

xii. Condensers (air, water cooled, spiral)

xiii. Reagent bottles (colourless and coloured, etched, plain) (250, 500, 1000, 2000, 5000 ml)

xiv. Aspirator bottles (2.5, 5, 10 liters)

xv. Desiccators

xvi. Bell jars

xvii. Adopters of different sizes.

xviii. Splash heads

xix. Soxhlet's apparatus with extractors of different capacities

xx. Capillary tubes and glass rods of different thickness and lengths.

The above list is not exhaustive but contains many of the common glasswares found and used in a laboratory.

e. Distillation apparatus : This is one of the important glass equipments used in any analytical laboratory for distillation work. It consists of a distillation flask (500, 1000 or 1500 ml capacity) fitted with one or two holed rubber cork. A condenser is fitted to one of the holes in the rubber cork of the distillation flask through a bent tube. The other end of the condenser is connected by a rubber tubing to a delivery tube (glass). The end of the delivery tube should dip into the solution kept in container (usually a beaker) but should not touch the bottom of the container. If it touches the bottom, then there is always a possibility of pressure building up in the distillation set and the whole set blowing off. At the same time the delivery tube should also not be above the surface of the liquid in the beaker as this will result in loss of gas or volatile liquids without getting absorbed in the liquid.

Fluctuations of gas pressure is a common experience in any laboratory. Once the distillation has been initiated, the amount of heat should never be reduced or cut off as this will result in back suction of the solution kept in beaker or ice tumbler at the delivery end of the condenser. When such a fall of pressure or heat is noticed, the delivery tube should be disconnected immediately, allowed to remain for a minute or two and then reconnected. If back suction has already occurred or suspected to have occurred, it is better to repeat the experiment.

To avoid such fall of gas pressure and also overcome the risks due to the presence of naked flames, nowadays heating mantles which are electrically operated and capable of maintaining constant temperature are used. They are available in different sizes to suit various purposes. Though they are very convenient and safe, still they are very costly and many laboratories particularly the teaching institutions cannot afford to use them for regular work. One advantage of these heating mantles is that even in the event of current failure, the heat will be maintained for some time and during that period, any corrective steps that are needed to overcome back suction can be taken and thus avoid loss of time and energy.

Another important part of distillation set is the condenser used for condensing vapours into liquid condition. There are different types of condensers like air condensers, water cooled condensers of different patterns. The air condensers are normally fitted only to digestion flasks and they are long, narrow glass tubes of 1 to 2 meters long with 2 to 5 mm diameter bore. In this the hot vapours move up and get cooled due to lower temperature of the surrounding atmosphere and come down to the digestion flask. In the process of flow back, they cool further lot of vapours and this goes on. However it is not very efficient one as in many cases, particularly when boiling is rapid, vapours escape without condensing. This can be avoided by using a very long tube but it is not practicable as invariably the whole assembly tilts and breaks even with a slight wind or touch of anything. Hence a compromise between efficiency and practical utility has to be made in this case.

The above problem has been overcome in the case of water cooled condensers. This type of condensers consist of long slender tube of 7 to 15 mm internal diameter of 30 to 75 cm length fused to an outer jacket of a glass tube with two holes - one for inlet and the other for outlet for water. Through the outer jacket, water can be allowed to move at a constant rate, usually the hole at the lower end of the condenser (inlet) is connected to a tap and water is allowed to gradually rise in the condenser and flow through the other hole at the top (outlet). The water that is circulating in the outer jacket absorbs the heat of the inside vapours and moves up and is removed at the outlet. One of the common mistakes committed in connecting the water source is to reverse the positions of inlet and outlet, that is to connect the top of the condenser to the water source and have the bottom as outlet. In this arrangement, the water just flows along the sides of the outer tube without touching the inner tube and also not giving any time for absorbing the heat. This will result in poor condensation of the vapours.

As the length of the inside condensing tube required for efficient condensation is invariably long and also the whole assembly occupies considerable space, attempts have been made to reduce the length without sacrificing the efficiency. Nowadays spiral condensers have come to use wherein the long glass tube is bent in the form of spiral and fused inside a glass jacket through which water could be circulated. By this the length of the condenser tube can be increased significantly but at the same time kept inside a short compact outer jacket. In the spiral condensers, the vapours have to traverse a longer distance before reaching the delivery end and as such have to be in contact with the cooling system for a longer time. This will result in greater efficiency of condensation. The stability of the set is also more due to its compact size and hence breakages are less than the ordinary sets. However these are slightly costlier than the ordinary one.

f. Soxhlet apparatus: The soxhlet apparatus is used for "extraction" of certain constituents in a mixture by making use of partial solubility principle. The apparatus consists of three parts - extraction flask, extractor with the thimble and the condenser. A weighed sample is transferred to a porous thimble kept in the upper extractor flask. This flask is connected to a extraction flask containing a known volume of the solvent. When the extraction flask is warmed, the solvent vapourises, passes through the tube and condenses in the upper flask and remains in contact with the substance as warm solution. The level of condensing liquid goes on increasing with time and when it reaches the level of siphoning, it automatically gets siphoned off. During the period of contact, the desired constituent automatically gets extracted by the solvent, the siphoned liquid goes back to the extraction flask, again gets warmed up, gets converted into vapour, moves up and collects in the upper extractor flask. This process repeats itself automatically. Usually within 6 to 8 such siphonings all the desired constituents would have been extracted.

However in certain occasions due to either manufacturing defect or due to improper alignment automatic siphoning does not occur. In such cases, the apparatus may be slightly tilted to initiate the siphoning. It is better to watch and count the number of siphonings that have occurred than to assume them in view of the above problem. The soxhlet apparatus comes in different sizes to suit different purposes.

g. Desiccators: These are used to prevent any material from absorbing moisture from the atmosphere. The desiccator consists of lower round cup-like part and a glass lid. Both portions have smooth rim portions which finely match each other. Usually Vaseline or other soft non-contaminating material is applied to these parts and to the lid on its lower part. This will prevent any air from entering inside the desiccator. Invariably hot materials straight from the oven or from burners are kept inside the desiccators for cooling. The high temperature results in the expansion of the air and also melting of the Vaseline which makes the lid slip and fall down. As the lid is of thick glass and heavy, when it falls, it breaks and finding a suitable replacement has always been a problem. Added to it if the same is a part of a vacuum desiccator, then the cost of replacement will be very high. To prevent this and provide a slight rough surface, before keeping the lid above the lower part, a small piece of paper 5 mm wide and 2 to 3 cm long may be kept between the two parts and the lid slid to close the desiccator. This will prevent the slipping of the lid.

To keep the atmosphere dry inside the desiccator, some drying agents like calcium sulphate, calcium chloride, phosphorus pentoxide, magnesium perchlorate, sulphuric acid may be kept inside in the bottom of the lower half and on the top of it a suitable porcelain holed tile kept for keeping the samples. It is advisable not to keep too hot substances inside and wherever necessary replace the desiccants as and when they become too moist in the case of solids and too dilute in the case of liquid desiccants like sulphuric acid.

By keeping different concentrations of sulphuric acid, the desiccator can also be used for keeping samples at a particular relative humidity. Also by connecting it to a vacuum source, the same apparatus could be used for drying in vacuum.

h. Other basic equipments :

i. *Wiley mill* : This is a common grinding mill used for grinding dry plant material. In this the dried plant material is fed into the hopper and the ground material collected at the other end after pulverisation. For proper grinding, the plant material should have been well dried, friable and contain least amount of moisture. More the amount of moisture, more is lump formation and also uneven outflow of materials.

As plant materials from different sources and of different treatments are received for grinding, the inside of the wiley mill should be cleaned immediately after grinding each sample. Also if the samples are to be taken up for micronutrient analysis like Cu, Fe, Zn, Pb, then wiley mill fitted with stainless blades alone should be used.

ii. *Water bath*: In many experiments the volume of filtrates or extracts of samples are voluminous and the solutions are too dilute for any titration. Hence they have to be concentrated to manageable volume. Also in certain cases, either precipitation process have to be hastened or process of formation of precipitation increased so that the particles are larger in size and do not pass through the filter paper or filter beds. Also the extracts in certain cases have to be dried for estimation of total soluble salts content. For these purposes a water bath is generally used.

Water baths are available in many sizes and are either electrically operated or heated with gas burners. When they are electrically operated, care should be taken to see that the water level never goes below the level of heating coil. The water level as well as the water temperature should be frequently checked and this should be done only after switching off the current supply as otherwise there is a possibility of getting electric shocks if the heating coil is old and defective. The maximum temperature attained is around 150°C. Any apparatus that is dried over a water bath is likely to have moisture and salts condensed on the outer side of the containers. Hence it is better to wipe the outer side of the container and clean it before weighing and also use low salt water or rain water in the water bath.

iii. Sand bath : This is used for digestion of materials with solvents. A sand bath is a metal sheet, usually tin sheet, folded on all sides in such a way that it can hold atleast 2 to 4 cm thickness of sand on it. This is also available in the market just like the water bath. This is either electrically heated or heated by bunsen burner. The rate of evaporation of the solvents or liquids kept on the sand bath is comparatively rapid and as such, the contents may become dry within a short time and the containers may even break. Hence the conical flask or other containers are either fitted with an air condenser or covered with a watch glass or funnel to reduce the rate of evaporation and at the same time allow the material to get digested. Commercially sand baths of various sizes are available but they can be manufactured even locally with the tin sheets.

iv. *Air oven* : To dry the containers or plant materials or any other material at constant temperature but at the same time without contamination, an air oven is usually preferred. In this hot air is allowed to move gradually from the bottom of the air oven to the top and in this process, it is made to absorb moisture from the materials kept inside and dry them. As most of the air ovens have thermostat controls, the required temperature may be attained and maintained for a long time. There is likely to be slight variation in the inside temperature at different points and invariably it is slightly higher at the entrance of hot air into the air oven and less near the exit. Most of the present day air ovens are electrically operated and hence care has to be exercised to see that the materials kept inside do not dry to the extent of catching fire. Whenever smell of burnt materials is felt, it is better to check up the air oven as otherwise the whole thing may catch fire. Invariably materials like filter paper and plant materials get charred when kept for a long time and also when the temperature rises suddenly inside the oven due to some disturbance in the circuit.

v. *Hot plate* : To digest materials at constant temperature and rapidly, a hot plate may be used. This is similar to a heater with a flat top and a thermostat control. As temperatures may rise suddenly, evaporation or digestion or boiling may become rapid and violent and may result in loss of materials. Hence extra care has to be bestowed while digesting the materials on a hot plate.

vi. *Incubator* : To maintain constant temperature usually lower than 80°C, incubators are used. Now a days these are electrically operated and have a thermostat control to maintain the desired temperatures. Normally incubators are not used in chemical laboratories except when the analytical work is linked with microbiological studies.

vii. *Shakers* : Most of the estimations of soils require constant shaking over a period of time and this cannot be performed manually. Hence electrically operated shakers are used. They are available in different sizes, shapes and patterns. The important one is the end over end shaker wherein the bottles containing suspension are fitted. Great care has to be exercised in fitting the bottles in the shaker. As these are generally fitted in two rows, the number of samples that are fitted should be even. In case odd number of samples are to be analysed, then a dummy container with equal weight of water may be added as additional sample and fitted so that the shaker is balanced on both sides. If this is not done, there is likelihood of jerking and also reducing the life span of the shaker.

Another pattern used for shaking small volumes of suspensions continuously is the wrist action or horizontal type shakers. Here the movement of shaking is horizontal. The advantage is that large number of samples can be handled at a time and the table space occupied by them and the shaker is also small. At the same time if the amount of suspensions in each container is more, then the efficiency of mixing will be reduced as most of the solid materials will remain at the bottom and the supernatant liquid alone will be shaken.

viii. *Centrifuges :* For separation of colloidal particles from a suspension and also in cases when the particles are so fine that they pass through the filter paper, centrifuging of the suspensions can be resorted. Centrifuges come in different sizes and with different speeds. The inside centrifuge cups are usually made of a metal and are even in number. Inside these cups there is always a rubber mat or cushion to act as a buffer for shock. The centrifuge tubes may be of glass or plastic. Whenever centrifuges have to be operated, the number of samples kept inside the equipment should be even and balanced evenly on both sides. In case odd number of samples have to be analysed, dummy tubes or bottles of equal weights may be added and centrifuged.

To prevent damage to the centrifuge and also loss of valuable experimental materials during centrifuging, the following precautions may be observed :

a. The tubes should be evenly balanced. Opposite tubes should be of the same weight. This is very important in high speed centrifuges working at more than 1000 rpm.

b. The speed of revolutions should be gradually increased and attained only slowly.

c. As far as possible, corrosive liquids like concentrated acids and alkalies should not be centrifuged.

d. Though most of the high speed centrifuges are fitted with timers, still it is preferable to check the centrifuge at the end of the required time.

ix. *Muffle furnaces* : In many laboratories in recent years, muffle furnaces are being extensively used for ashing the plant sample and also for removing moisture at high temperatures like 500°C. Each muffle furnace has an energy regulator which gives step control over a wide range and an indicator lamp. The temperature of the furnace can be raised to even as high as 1500°C (in certain cases even upto 2000°C) and maintained for a long time by means of blocks of heat resistant bricks used for the walls of the muffle furnaces. The actual temperature inside is measured with a calomel - alumol thermocouple and is indicated on a meter outside. Also when the temperature rises above the limits, the equipments will go off by blowing the inbuilt fuse. As these equipments are very heavy, it has to be kept in a place with strong base. Also as high voltage of current passes and the muffle furnace also invariably operated most of the days, it should be connected to a separate line and not to lines where other electrical equipments operate. Though well insulated, still the temperature of the air around the muffle furnace is always higher than the room temperature and hence equipments that are likely to be affected by the higher temperature should not be kept near by. Similarly balances should also not be kept near a muffle furnace.

x. *Other apparatus* : In addition to the apparatus and equipments listed above, the following apparatus are also commonly needed for the various analysis.

a. Silica crucible (10, 20, 50 ml)

b. Silica basin (5, 7.5, 10, 12.5 cm diameter)

c. Porcelain basins (5, 7.5, 10, 15 cm diameter)

d. Potash basins (7.5, 10 cm diameter)

e. Porcelain tiles

f. Porcelain crucibles for filtration (10, 25, 50 ml)

g. Tripod stands, filter stands, burettes stands, retort rings, retort stands, sampling knives, sieves, rubber corks of various sizes, borer sets, wire gauge, wooden mallet, wide mouth sampling bottles, asbestos mat, thermometers.

h. Other costly equipments, like solu-bridge, pH meters, colorimeters, flame photometers, Atomic Absorption Spectrophotometers, spectrophotometers (the important principles on which these work are detailed in the next chapter).

CHAPTER - 7

PRINCIPLES IN INSTRUMENTAL METHODS OF ANALYSIS

The instrumental methods of analysis involves the use of instruments, simple as well as sophisticated, of varying degrees of accuracy and sensitivity in the estimation of elements or compounds. The accuracy of estimation mainly depends on the sensitivity and sophistication of the equipments used. Greater the accuracy and precision of the instrument, the greater the accuracy that can be expected in an estimation. At the same time greater sophistication increases the cost of the equipments. Hence in the final choice of an instrument, consideration has to be given to the initial cost, flexibility of operation and adoption, and also special needs of the concerned laboratory.

The instrumental methods can be grouped under various heads based on the principles used in potentiometric method. The principles of Wheatstone's net work is adopted while in the conductometric methods the conductance of current is measured. In polarimeter methods, polarisation of ions or molecules of the particular groups or elements are measured. Various techniques used to measure the intensity of colour of the solution come under colorimetric methods. Measurement of emission of light of an element or particular compound is done using a flame photometer. In the following pages the important principles of the methods used in the analysis of materials referred to in this book are discussed briefly.

A. COLORIMETRY

Colour of a precipitate or a compound has been used as a criteria for identifying the elements or radicals or compounds in the qualitative analysis. Now the same principles have been adopted with some modifications in the methods to quantify the amount of the constituent present in microquantities. With the enunciation of Lambert-Beer's law and also the development of photoelectric cells, number of techniques have been developed in the recent past to such an extent that a separate branch of analysis called COLORIMETRY has been added to the quantitative analysis.

The variations in the intensity of the colour of a system due to the concentration of some component forms the basis of colorimetry. The colour is due to the inherent nature of the constituent itself or due to the formation of a coloured compound by the addition of an appropriate reagent. The intensity of the colour in the test solution can be compared with that observed by treating a known amount of the substance in the same manner and already quantified. A coloured solution may sometimes be formed as soon as the substance is dissolved in the solvent and ions such as MnO_4, CrO_4 having sufficiently intense colour are produced or a coloured solution may be produced by the addition of some reagent which reacts chemically with the element or ion under test as in the determination of Fe by the addition of ammonium thiocyanate. In cases where the colour of the solution of salts is not intense enough to measure as in the case of copper salts, they are converted into an intensely azure-blue coloured complex $\{Cu(NH_3)_4\}^{2+}$ by the addition of ammonia. In some other cases, there may exist an equilibrium between the different coloured compounds in the same solution. In such cases the equilibrium has to be displaced by some means to give a definite determinable form of intermediate shade of colour to the solution which can be determined. This occurs in the determination of compounds at relatively low concentrations.

In the colorimetric and spectrophotometric methods, only solutions that absorb light can be used. The accuracy and sensitivity of the determinations will be greater when the light of greater intensity is absorbed. The concentrations and the layer thickness of the solution under consideration being constant, its optical density is determined by the molar extinction coefficients which is a measure of the sensitivity of the colorimetric method.

When white light is passed through any coloured solution,

the solution preferentially absorbs some wavelengths and the colour observed is characteristic of wavelengths that have not been absorbed as seen from the table below;

ABSORBED RADIATION		Hue of the colour transmitted (observed)
Hue of the colour	Wavelength in mμ	
Violet	400 – 450	Yellowish green
Blue	450 – 480	Yellow
Greenish blue	480 – 490	Orange
Bluish green	490 – 500	Red
Green	500 – 560	Purple
Yellowish green	560 – 575	Violet
Yellow	575 – 590	Blue
Orange	590 – 625	Greenish blue
Red	625 – 750	Bluish green

Hence in colorimetric analysis, a band of wavelengths is selected and then it is allowed to pass through the solution and the intensity of the emergent light or the ratio of the intensities of the incident and emergent light is studied. The ideal band or wavelength is selected by constructing the absorption curve for the visible spectrum by means of a spectrophotometer. Comparison of the curve with the spectral transmission curves of the set of filters supplied by the manufacturers help in making suitable choice. Alternatively absorbance or transmission-concentration calibration curves may be constructed with a photoelectric colorimeter using each one of the filters in turn. The best filter to use in a particular determination will be one which gives the maximum absorbance or minimum transmission for a given concentration of the absorbing substance. The other methods where the filter that gives the least transmission for a given concentration and depth of the cell is also used. Choice of filters which are close to the complementary colours of the solution is less satisfactory but can be used if there are no other alternatives. The colour or the wavelength band of the incident light used is selected to coincide with the colour most absorbed by the test solution and this is done by the passage of light through a light filter.

When light (monochromatic or heterogeneous) falls on a homogeneous medium, a portion of the incident light is reflected, another portion is absorbed within the medium and remainder is transmitted. This could be mathematically expressed as

$$I_o = I_a + I_t + I_r$$

where

I_o = Intensity of incident light

I_a = Intensity of light absorbed

I_t = Intensity of light transmitted

I_r = Intensity of reflected light

For air - glass interference (where glass cells are used) about 4% of the incident light is reflected. It is usually eliminated by the use of a comparison cell. Then the above equation is reduced to

$$I_o = I_a + I_t$$

The relationship between these three variables varies with the optical properties of a particular solution i.e. its stability to absorb and transmit light. This relationship is used in colorimetry wherein the quantity of a coloured constituent is determined by measuring the relative amount of absorption of light by optical density (OD) or transmittance (T) with reference to the known concentration of the substance.

Colorimetry is based on two important laws namely Lambert's law and Beer's law. These two have been combined and is being called "LAMBERT-BEER'S LAW". Lambert law (1760) states that when monochromatic light passes through a transparent medium, the rate of decrease in intensity with the thickness of the medium is proportional to the light intensity. In other words, the intensity of emitted light decreases exponentially as the thickness of the absorbing medium increases arithmetically, indicating that each layer of given thickness of the medium absorbs the same fraction of the incident light falling upon it. For example, if one centimeter thick of a medium absorbs one tenth of the intensity of radiation then the reduction in intensity due to the thicker layers will be as follows:

Thickness of layer cm	Reduction in intensity					Fraction of intensity transmitted
1 cm	0.1					0.9
2 cm	0.1	+	0.09	=	0.19	0.81
3 cm	0.19	+	0.081	=	0.271	0.729
4 cm	0.271	+	0.0729	=	0.344*	0.656
5 cm	0.344	+	0.0656	=	0.410*	0.590
6 cm	0.410	+	0.0590	=	0.469	0.531

*rounded to the nearest next numbers

From this it could be seen that for a 6 cm thickness the reduction in intensity is only 0.469 and not 0.6 of the incident light.

Beer (1852) studied the effect of concentration of the coloured constituent in a solution upon the light transmitted or absorbed and found that the intensity of a beam of monochromatic light decreases exponentially as the concentration of the absorbing substances increase.

Combining both these laws, it can be stated that the rate of decrease in the intensity of radiation with the thickness of the absorbing solution is proportional not only to the intensity of the incident radiation but also to the concentration of the solution. Expressed mathematically it is

$$I_t = I_o \times 10^{-\varepsilon Cl} \text{ or } \mathrm{Log}\,(I_o/I_t) = \varepsilon\, Cl = A = od$$

where

I_o = Intensity of incident light

I_t = Intensity of transmitted light

ε = Molar extinction coefficient (molar absorptivity, molar absorption coefficient)

C = Concentration of the solution (moles/liter)

l = Length of the absorbing solution (cm)

A = Absorbance or optical density (od)

The value of ε (molar extinction coefficient) depends upon the units used for concentration and length. It is actually the optical density of the solution in which the concentration of the coloured substance is 1 mole /liter, the thickness of the absorbing layer 1 cm. It depends on the properties of the solute and on the power to absorb light.

The common logarithm of the ratio of the intensity of incident light to the intensity of transmitted light is known as "OPTICAL DENSITY" or "ABSORBANCE OF THE SOLUTION" and expressed as $D = \mathrm{Log}(I_o/I_t)$ indicating that the optical density of a solution is directly proportional to the concentration of the solution and thickness of the absorbing layer. The ratio I_t/I_o is the fraction of the incident light transmitted by a thickness "t" of the medium and is termed as TRANSMISSION or TRANSMITTANCE (T). It is the reciprocal of the capacity and the optical density of the medium. Thus a medium with optical density D for a given wavelength transmits 10% of the incident light at that wavelength. The absorbance "A" sometimes called optical density is defined as the logarithm to the base 10 of the reciprocal of the transmittance. $A = \log_{10}$ (I/T). Absorbance and transmittance have no units and are stated as a decimal or in per cent.

The Lambert - Beer's law also called Bougner-Lambert-Beer law holds good under the following conditions :

i. Light of a single wavelength alone is used (i.e. monochromatic light).

ii. The solution should contain only one species capable of absorbing the particular wavelength (a species is a single form of an ion or molecule. An ion with five molecules of water coordinated is a different species compared to the same ion with four molecules of water coordinated).

iii. Only the light transmitted through the solution is measured as emergent light and not the fluorescence or scattered light by the suspended particles or leakage by the instrument.

iv. Only dilute solutions are used. Discrepancies will arise when the solutions are concentrated or when they undergo association or dissociation.

Instrumental factors or interactions within the solutions introduce errors and they have to be eliminated. Some of the instrumental errors are:

i. *Too wide a slit*: Monochromatic light can be obtained only at infinitely narrow slit width. When the slit is wide, then the condition is not satisfied. But too narrow a width will not produce sufficient signal reading for the detector to detect compared to the background noise. Hence in practice, a width size not too wide but at the same time wide enough to produce signals of the desired minimum level has to be adopted.

ii. *Stray radiation*: Scattered radiation from the end point of the spectrum and that is not absorbed by the sample may fall on the detector. This may give a negative deviation and the effects can be very serious when the spectrometer is used close to the limits of its range.

iii. *Reflection losses*: Reflection from the front and back surfaces of the windows results in losses in the path of radiation beams. No two cells give the same reflection losses and the losses increase as the optical windows deteriorate due to scratching and wear and tear. Hence great care should be taken in handling them.

iv. *Path length errors*: If the cells are not placed perfectly normal to the path of the beam, the path length traversed through the solution will be greater than the measured cell thickness and this may cause deviations. Though each of the tubes supplied with the instrument is expected to be uniform in whatever direction it is kept, still there is a possibility of some minute variation in the size and composition of the tubes and the material even within the same tube. Hence many manufacturers supply the tubes marked with a white vertical line so that the tube may be placed in the same place always and also in such a manner that the white line coincides with a particular point on the equipment. This will overcome much of the path length errors.

v. *Solution interactions:* If the solution is not homogeneous and is found coagulated into a small number of large units, the area through which radiation passes will be different. Also secondary scattering of the beam would cause an opposite effect and the net overall effect will be complicated and difficult to quantify. Hence errors due to coagulation should be avoided as far as possible.

vi. *Solute-solvate interactions* : Changes in concentration may cause changes in the nature of absorbing species due to molecular self association and this will be more in compounds with H-bonding. This would cause deviations also. This could be overcome by working all samples in a short definite range of concentrations.

vii. *Solute - solvent interactions*: Similar to the above, the association of solute - solvent may also cause deviations.

viii. *Fluorescence* : Strong fluorescence effects similar to stray light will also induce deviations.

ix. *Chemical side reactions* : Altering the experimental conditions like pH, temperature, time lag between mixing and estimations, addition of complexing agents alter the concentration as well as properties of the different absorbing species present in the solution, resulting in deviations. Hence in analysing every batch of samples uniformly the same procedure should be followed. Also the time factor should always be carefully looked into as this is the major source for deviation in most of the analysis.

The errors due to the above factors could be overcome partly or completely by following rigid experimental conditions during the colorimetric analysis to the extent possible. The following guidelines will help to overcome the errors to a large extent:

i. *Specificity of a colour reactions*: Though very few reactions are specific for a particular compound, still by maintaining certain conditions like correct oxidation states, pH, temperature and addition of certain specific complexing agents wherever possible, the required specificity could be attained.

ii. *Proportionality between colour and concentration* : A linear relationship between colour and concentration should be maintained in the visual colorimetric estimations. In the photoelectric estimations, this may not be necessary as a calibration curve is generally drawn relating the colour to the concentration of the solution. But still in any case, the relationship should follow the Lambert-Beer's law.

iii. *Stability of the colours* : This is a very important factor which introduces errors in estimations. Most solutions reach maximum colour intensity after a time and thereafter fade out. The procedure followed should be such that the period of maximum colour intensity is long enough to prepare other samples and also for making precise measurements of the samples already prepared.

iv. *Reproducibility* : The choice of procedures should be such that they could be reproduced with the same degree of precision and accuracy under identical conditions.

v. *Clarity of the solutions* : The solutions prepared for the colorimetric estimations should be clear and not turbid or coagulated.

vi. *High sensitivity* : As the colorimetric estimations are adopted for estimations of constituents present in low amounts, many a time they are present in parts per million (ppm) level, the colour reactions should be highly sensitive. It is desirable that the reaction products absorb the light in the visible range than in the ultraviolet range as the influence of interfering substances in the ultraviolet range is usually more pronounced.

Inspite of the above limitations and rigid conditions, the colorimetric methods often give more accurate results at low concentrations than the corresponding volumetric or gravimetric procedures. The colorimetric estimations are simple and can be frequently applied when no satisfactory gravimetric or volumetric procedures are available for certain biological estimations. Further these procedures can be adopted for routine determinations of compounds of a number of similar samples by virtue of rapidity with which they can be performed and at the same time with the greatest accuracy when conditions are rigidly controlled. The chief advantage of these methods is that they provide a simple means for determining minute quantities of substances. The upper limit of colorimetric methods is that the constituent in the sample should be present in amounts less than 2%.

These analytical methods involve the estimations of the quantity of the constituent by measuring the degree of absorption of light of certain wavelengths by coloured substance. Earlier all such measurements were made visually and was only approximate and also the degree of accuracy depended upon the experimenter's capacity to distinguish different degrees of hues. This has been overcome now with the use of photoelectric cells and the visual methods of colorimetry are becoming obsolete.

In visual colorimetry, natural or artificial white light is generally used as a light source and determinations are usually made with a single instrument called colorimeter and colour comparator. Here the human eye matches the intensities of colour as the analyst cannot remember the different hues or intensities. The following visual colorimetric methods were adopted earlier for the estimations:

i. *Standard series method* : In this a series of Nessler's tubes are filled to the mark with solutions containing various known quantities of the desired constituent. The unknown is prepared and a comparison of its colour is made with the colour of the known solution. For this, the tube containing the unknown is matched with the standard already made and when any one standard exactly matches, then the concentration is taken to be same as that of the standard. In case the colour of the unknown is found to lie in between the colours of two standards, a new series of standards more closely matching the unknown is prepared and matched or the concentration of the unknown is approximately fixed. Coloured glass standards may also be substituted for the standard solutions wherever their use is sufficient. The error in this method is of the order of ±8%. For convenience artificial standards like Loviband glasses, salt solutions such as ferric chloride in aqueous hydrogen chloride (yellow), aqueous cobalt chloride (pink), aqueous copper sulphate (blue), aqueous potassium chromate (orange) may also be used but all these have to be standardised earlier. The main disadvantage of this method is that the spectral absorption curves of the test solutions and of the standard glasses or solutions may be far from identical. Further the error will be increased if the observer is suffering from partial blindness.

ii. *Duplication methods* : A standard solution of the constituent under test is added to the reagent until the colour produced matches that of the unknown in the same volume of the solution. This is less accurate than the standard series method.

iii. *Dilution method* : The sample and the standard solutions are kept in glass tubes of the same diameter and are observed horizontally through the tubes. The more concentrated solution is diluted until the colours are identical in intensity when observed horizontally through the same thickness of the solution. The relative concentration of the original solution is proportional to the height of the matched solutions in the tubes. This is still less accurate.

iv. *Balancing method* : This is the basis of all colorimeters of the plunger type (Duboseq colorimeter). Comparison is made in two tubes and the heights of the liquid in one tube is adjusted so that when both tubes are observed vertically, the colour intensities in the tubes are equal. The concentration in one of the tubes being known, then that in the other may be calculated from the respective lengths of the two coloured liquids using the relationships $C_1t_1 = C_2t_2$ where t_1 and t_2 are the lengths of the columns and C_1 and C_2 are the concentrations of the solutions. As white light is usually employed, the accuracy obtainable in a Duboseq colorimeter is of the order of $\pm$ 7%. The accuracy can be improved appreciably if monochromatic light is used.

In a Duboseq colorimeter, the transmitted light from the two solutions are placed side by side so that they could be viewed by a single eye. Any difference in hue or intensity would cause a line of demarcation between the two halves of the disc.

However there are number of disadvantages in the visual colorimetric methods. Some of them are:

i. A standard or series of standards must always be available.

ii. The eye is not capable of matching the solutions which contain a second coloured substance in it.

iii. The eye is not as sensitive to small differences in absorbance as the photoelectric devices. As a result concentration differences smaller than about 5%, relative to the other one cannot be detected.

iv. In view of the varying observing capacities of the individuals, different conclusions may be reported by different analysts on the same colour.

Hence observations by human eye is being replaced by a suitable photoelectric cell which offers direct measure of the light intensity and hence the absorption and not the colour of the substance. For this reason, the term photoelectric colorimeter is a misnomer. Better names are photoelectric comparator, photoelectric photometer or best absorptiometer. The visual methods are gradually being displaced by methods using photoelectric cells (filter photometers) or absorptiometers and spectrophotometers, thus reducing the experimental errors in colorimetric determinations. The instruments used are called filter photometers and spectrophotometers. They consist of a light source, a suitable light filter to provide an approximate monochromatic light, a glass cell for the solution, a photoelectric cell to receive the radiation transmitted through the solution and a measuring device to determine the response of the photoelectric cell. The photoelectric colorimeter operates with a light of a comparatively narrow range of wavelength obtained by passing white light through materials present in the form of plates of coloured glass, gelatin etc. transmitting only a limited spectral region. The comparator is first calibrated in terms of series of solutions of known concentration and the values plotted on a graph with the concentrations in the X-axis and readings in the Y-axis. With these points a curve is drawn. The concentration of the unknown solution is then determined by noting the reading in the meter and then referring to the standard curves.

The photoelectric colorimeters (absorptiometers) can be divided into two main classes viz. one cell (single beam) and two cell (double beam) instruments. In a direct reading photoelectric photometer, the light passing through the sample from a source of constant illumination is made to impinge on the photovoltaic cell and the electric impulses are registered in a galvanometer and it gives a measure of relative intensity. A filter is used to give a monochromatic light.

In a single beam direct reading colorimeter the optical path is simply from the light source through the filter, sample holder and to the detector (spot galvanometer) eg. EEL colorimeter, Hilgard, Biochem, Evelyn photoelectric colorimeter, Immetron colorimeter 450 etc.

The single beam colorimeter consists of a tungsten filament bulb, a lens to provide a parallel beam of light, a filter to restrict the light to desired beam wavelength and a photovoltaic cell. The current produced in the photovoltaic cell is measured with a micrometer. Almost in all the single beam instruments, provisions is made for reading directly in terms of per cent transmittance. The linear scale has a graduation of 0 to 100. A cell containing only the solvent is first placed in the light path and the power of the beam adjusted until the needle indicates zero or 100 as the case may be. It is done by adjusting the opening of the diaphragm to allow the light path according to the requirement. The sample is then placed in the path of the beam. Since the signal from photovoltaic cell is linear, the resulting scale reading will give the per cent transmission while the logaritham scale would give the absorbance by the solution directly. The main defect in the single beam colorimeter is that the fluctuations in the light intensity affects the measurements of transmittance and there is no mechanism to indicate the degree of fluctuations and make any correction. This has been overcome in the double beam colorimeter where only the difference between the sample and reference beam energies is obtained as a signal and measured.

Double beam colorimeters (Klett - Summerson photoelectric colorimeter, Spectronic 20 (Baush and Lomb), Fisher electrophotometer, Immetron photoelectric colorimeter model 402 etc); In the double beam colorimeters, the light beam is divided into two, part of it passing through the solution in the cuvette before falling on the measuring photocell and the other part passing directly on to the reference photocell that continuously monitors the output of the lamp. The output of the working photocell is then compared with that of the reference by a suitable circuit design. In this type, the current flows from the photovoltaic cells and passes through the variable resistance. One of these is calibrated on a transmittance in linear units from 0 to 100. A sensitive galvanometer serves as a null indicator and is connected across the two resistances. When the potential drop across one is equal to that across other, no current will pass through the galvanometer. Under all other circumstances, a current flow is observed.

Selection of filters for photometric analysis

Photometers are usually supplied with an array or set of filters, each one of which transmits a different portion of the spectrum. The selection of the proper filter to suit the given solution is very essential for correct assessment of the content of the constituent. The sensitivity of the measurements is directly dependent upon the filters. The most suitable filter can easily be selected according to the colour complement of the solution.

Spectronic 20 (Bausch and Lomb) is now widely used and can

be considered as representing a double beam colorimeter in which a photometric accuracy is sacrificed in return for simplicity and low cost. Its normal range is 350 to 690 mμ with a blue sensitive phototube and this range can be extended upto 950 mμ by the use of a red sensitive phototube and red filter. The monochromator system consists of a reflector grating, lenses and a pair of fixed slits. Because the grating produces a dispersion that is independent of wavelength, a constant band width of 20 mμ is obtained through the entire operating region. A single phototube is used in Spectronic 20. The output is amplified and used to actuate the pointer of a meter calibrated in terms of both transmittance and absorbance. Spectronic 20 is a direct grating spectrophotometer. The scale is graduated from 0 to 100 per cent transmittance of the light incident on the phototube. The instrument is balanced at 100 per cent transmittance by means of a variable 'V' shaped aperture. Also in this instrument the wavelength setting is of added advantage.

When the instrument is used for colorimetric applications, the wavelengths' control is normally set to a single wavelength in millimicron by means of a wavelength scale which is an added advantage. For accurate results, the index line of the matched test tubes should be set opposite the index line on the adopter. Once the light control has been set and the meter reads 100%, hundreds of samples can be run. The mirror on the scale is provided so that the position of the needle may be read more accurately.

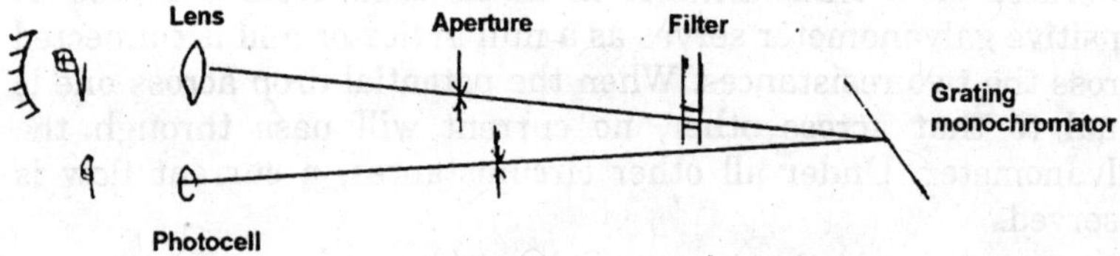

DOUBLE BEAM COLORIMETER

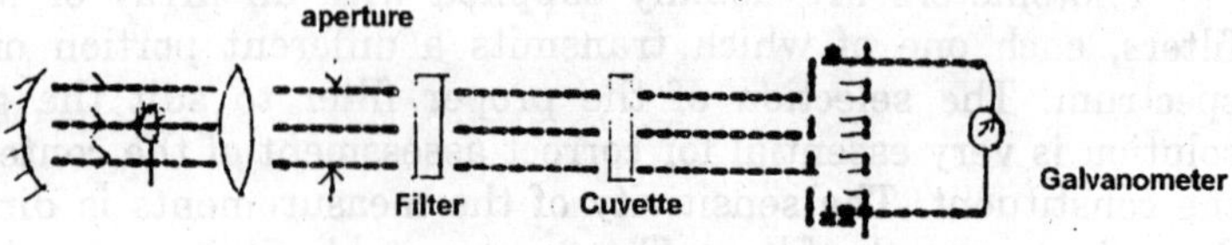

Single beam colorimeter

SPECTROPHOTOMETER

Both colorimetry and spectrophotometry are based on the Lambert-Beer's law. Spectrophotometry measures the relative amount of light transmitted by an absorbing medium. Spectrophotometric methods are more accurate for determining the concentration of the substances in solution than colorimetric techniques. Though variations of the colour of the system with change in the concentration of some component form the basis for both the methods, the colorimetric method is usually employed with light of a comparatively narrow range of wavelength obtained by passing the whole light through filters whereas in the spectrophotometric methods the following regions of the spectrum is used; UV - 180 - 400 mμ; Visible 400 - 760 mμ or 760 mμ - 25 μ. Spectrophotometry differs from colorimetry in that the absorbance of a solution or substance is determined with only one wavelength of light being transmitted and measured. In practice a narrow band of wavelength of light is used rather a single wavelength. Spectrophotometers can be regarded as refined filter photoelectric photometers permitting the use of continuously variable and more nearly monochromatic bands of light. In the spectrophotometric analysis, light of wavelength not exceeding 1 to 10 Å in band width extending to the UV region of the spectrum is used as a source of light.

A spectrophotometer is really made up of two instruments - a spectrometer and a photometer. A spectrometer produces coloured light of any selected colour or wavelength and when employed as a part of the spectrophotometer is usually termed as a monochromator and is calibrated in wavelength. A photometer measures the intensity of the light and in a spectrophotometer it is used to measure the intensity of the monochromatic beam produced by the associated monochromator. Generally photometric measurements are made first with a reference liquid and then with a coloured sample kept in a similar cell interposed in the light beam, the ratio of the two intensity measurements being a measure of the capacity of the same at the wavelength of the testing.

Colorimeters cannot be used to estimate concentrations of <0.001 μg as there will not be any visible colour development. But with a spectrophotometer even such low concentrations can be estimated. Added to this, there are less procedural steps in the estimations and this reduces the cumulative error. Also the methods are rapid and accurate and hence hundreds of samples could be analysed daily. However the spectrophotometric methods cannot be used for analysing elements present in macroquantities as the errors due dilutions will outweigh the other advantages. Usually this instrument is used for analysing samples containing 0.01 to 0.001 M concentration of the elements.

There are number of models of spectrophotometrs available in the market and they cover different ranges of wavelengths. For example Unicam SP 600 Spectrophotometer (visible and near IR region 300 - 1000 mμ), Hilger UV spectrophotometer (UV and visible), Unicam SP 500 Spectrophotometer (UV and visible), Beckman model B Spectrophotometer (UV and visible). The essential parts of a spectrophotometer are a source of radiant energy, a monochromator, glass or silica cell for the solvent and for the solution under test and a device for receiving or measuring the beam or beams of radiant energy passing through the solvent or solution. The versatile nature of this instrument is due to the availability of wide range of wavelengths. A tungsten filament incandescent bulb provides for the measurement in the visible region of the spectrum, and a hydrogen gas discharge tube for the UV range. Glowing rods of silicon carbide or zirconium oxides are used in the IR region. Glass is not transparent to UV rays. Hence the sample tubes (cells and cuvettes) are manufactured in quartz. The IR spectrum is absorbed by both glass and quartz. Hence prism and windows are fabricated from inorganic halogen salts such as NaCl or KBr.

The single beam - null balance spectrophotometer is a very accurate and sophisticated model than the earlier ones. The chosen sample after extraction is placed in the cuvette and light passed through it. When the beam penetrates the sample, energy is absorbed and the resultant light strikes the photocell. The amount of light transmitted is recorded and the percentages of the transmission is plotted against a range of wavelengths. The graph will indicate the peak from which identification can be made from the known value of the sample. Qualitative values can be obtained from the known maximum absorption wavelength. A standard curve for the specific compound will give the correct value. Correction is always done by setting the instrument to zero absorbance using a blank. Spectrophotometer working on mains are better than those capable of operating on both batteries and mains. The present models have both transmittance and absorbance scales. While double beam spectrophotometer are widely used for routine quantitative work at selected frequency, the highest accuracy can be obtained by a single beam operation using the manual method (cell in - cell out).

C. FLAME PHOTOMETRY

In qualitative analysis where a salt is introduced in a flame, the colour of the flame indicates the nature of cations present in the salts. For example a characteristic yellow colour is produced by sodium, red colour by calcium and violet light by potassium when these salts are introduced in a nonluminous flame. This property of emission of characteristic radiant by each element is made use of in flame photometry whereas the intensity of emission at a particular temperature of flame is measured and then reported in terms of concentration of the element in solution.

The principle in this technique is that when a metallic salt is introduced into a nonluminous flame, the salt is separated into its component atoms because of high temperature. The energy provided by the flame excites the atoms to higher energy levels and when the atoms return to the ground or unexcited state, they emit their characteristic radiations. Since the excitation can be to different levels, light of several wavelengths can be emitted. However the intensity of wavelength corresponding to the most probable transition will be the highest. Each individual atom emits one quantum of radiation. Therefore the intensity of radiation emitting from the flame will be proportional to the number of atoms in the flame, that is the concentration of the element in the flame. The concentration in turn is proportional to the concentration of the elements in that test solution.

If the liquid contains an element capable of excitation, the flame burns with the wavelength of the element and the light is then collected alone. By introduction of a suitable solution in the light path, all the regions of the spectrum except the characteristic wavelength or waveband, can be eliminated. The filtered light is then passed to a photocell which gets energised and records the intensity in a galvanometer.

As such the flame photometer consists of three important parts :

i. Atomisation assembly comprising of a nebuliser to produce a fine mist from the solution and a burner which vapourises the analyte.

ii. Monochromator system (filter, prism) which separates out the analytical wavelength from other radiations and

iii. Photometric system for measuring the intensity of emitted radiations.

The burner assembly used is usually of the type of "total consumption". The sample is aspirated into the flame by either the fuel or the oxidant but the gases are not premixed. The burner assembly is connected directly to the gas supply and the other to the compressed air through a spraying nozzle. The air stream carries with it to the burner flame a spray of the test solution. The gas and air pressure must be steady during the operation of the instrument and for this connections, regulators and burners should be checked before use. When large salt concentrations of high strength and acid solutions are fed into the burner, then the burners should be cleaned immediately after use.

Air at a given pressure is passed into air atomiser and the suction it produces draws a solution of the sample with the atmosphere when it joins the air stream in a fine mist and passes into the burner. Here in a small mixing chamber the air is mixed with the fuel gas supplied to the burner at a greater pressure and the mixture is burnt. The fraction of atoms excited depends upon the flame. If the temperature of the flame is increased the number of excited atoms also increase. Hence it is important to control the temperature of the flame between 1000□C to 3000□C so that the elements may be excited completely.

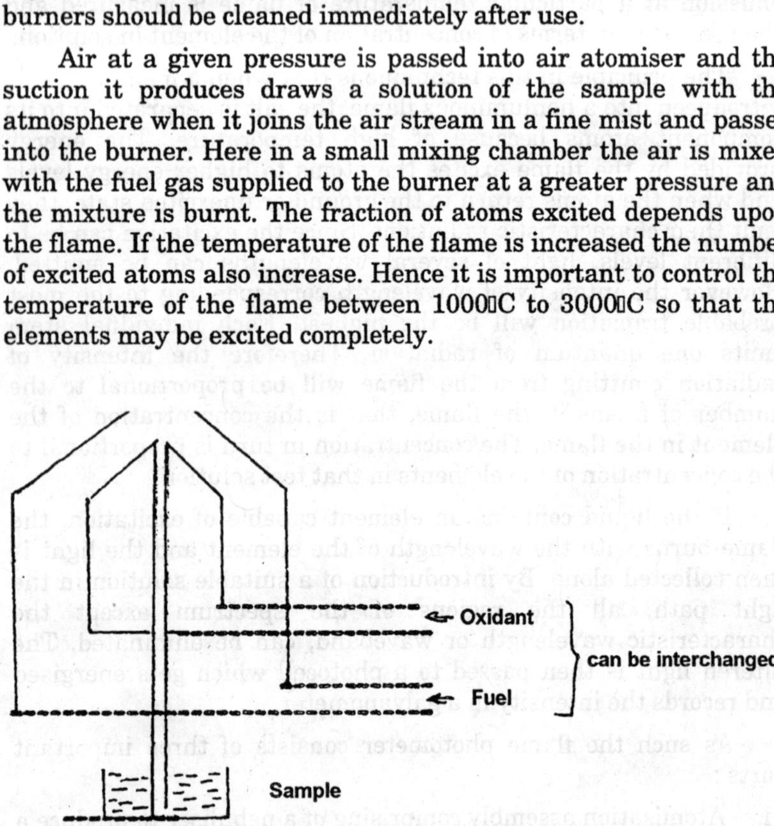

When the temperature of the flame is increased, the intensity of the emitted radiations generally is increased. But this is not true in the case of lithium, potassium and sodium which ionise easily. When these elements are heated to higher temperatures in the flame they get ionised instead of excited. Therefore such elements moderate temperatures of flame and produce only excited atoms. The flame temperature produced by different fuel-oxidant combinations are as follows :

Fuel	Flame temperature °C oxidant	
	Air	Oxygen
Propane	1900	2800
Hydrogen	2100	2780
Acetylene	2200	3050
Methane	2000	2700
Butane	1900	2900
Illuminating gas	1800	2800

Mixtures of coal gas and air does not give very hot flames because of the presence of nitrogen.

The burner should produce a steady flame and this flame should be able to decompose the compounds of the solid residue resulting in the formation of atoms. Also the flame should be able to evaporate the liquid droplets from the solutions and excite the atoms and make them emit radiant energy.

The temperature of the flame depends upon number of factors like amount of solvent entering the flame, type of fuel and oxidant, fuel to oxidant ratio, type of burner employed, and the region of the flame which is to be focussed at the entrance of the silt of the unit.

The radiations from the resulting flame is then passed through a lens, then through an iris diaphragm and finally through an optical filter which permits only radiations characteristic of the element under test to pass through to the photocell which measures the intensity of the radiation. The best detector is of the barrier layer cell type which has a high temperature coefficient. The output of the photocell is measured by a suitable galvanometer. The optical path from the chimney to the photocell is enclosed in the light box.

The filter interposed between the flame and the detector transmits only a strong line of the element. The simplest detector is a barrier layer cell, if sufficient energy reaches the cell. This cell has a high temperature coefficient and hence it must be placed at a cool part of the photometer. The precision can be improved by the use of internal standards, two filters and two phototubes (one for the standard and the other for the unknown). The basic assembly of the flame is as given in next page.

Flame photometry is a simple and rapid method for the estimation of elements which could be easily excited. It is also

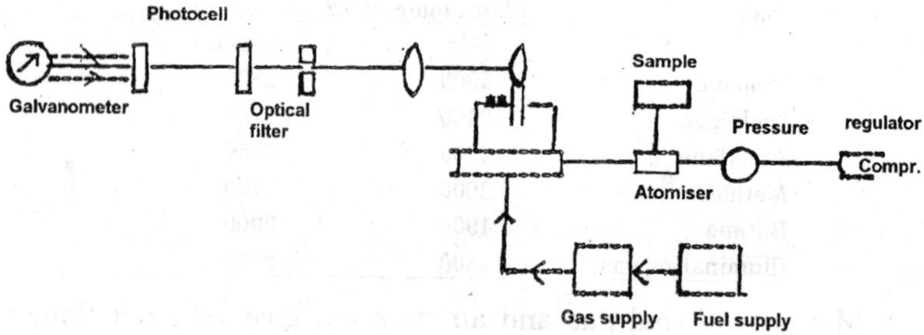

Flame photometer

known as flame emission spectroscopy as atoms are thermally excited in a flame. The flame must be hotter in atomic emission than for the same element in atomic absorption. By this technique even as low amount of 0.002 μg/gm of sodium or potassium can be detected. However in this technique inert gases and halides cannot be estimated. Also only liquid samples could be used. Further this method does not provide any information about the molecular structure of the compounds. Also the presence of other metals, viscosity of the solutions and presence of acid decrease the intensity. If temperatures are higher than that needed, the salts or their decomposition products vapourise and get dissociated. Then the readings are vitiated.

They are large number of flame photometers in the market but they can be grouped into two categories; i. single beam flame photometer and ii. double beam flame photometer. The single beam flame photometer consists of only one set of optics. Light from the centre of the flame just above the inner cone is collected by the reflector and focussed by a lens of heat resistant glass through the interchangeable optical filter on to single photodetector. Alternatively from the burner it passes into a monochromator and the radiation leaving the silt is focussed on to the detector.

The double beam flame photometer consists of a second light path provided for the radiation emitted by the internal standard elements (lithium) that is added in a fixed amount to each of the test solutions and calibration standard. The internal standards and the analytical wavelengths are isolated by means of optical filters in a dual optical system and each beam is focussed on to a separate photoelectric cell.

The commercially available flame photometers can be classified into four main groups :

i. Direct reading absorption filter barrier layer cell type eg. EEL corning models 2 and 3. Systeronics model, Elico model

ii. Internal standard - interference filter - barrier layer cell

iii. Prism monochromator direct reading or recording photomultiplier tube type eg. Beckman B-DB. Beckman model DV, Unicam Sp 900, Perkin- Elmer model 52C.

iv. Grating monochromator - reading photomultiplier tube model eg. Hitachi; Perkin - Elmer Model 139

ATOMIC ABSORPTION SPECTROSCOPY

Atomic absorption spectrophotometer (AAS)

This is a highly sophisticated, highly precise instrument capable of analysing as many as 73 elements when suitable hollow cathode lamps are used. The sample preparation is easy and the analysis can be carried even with samples containing as low as 0.001 ppm of the constituent and at the same time quickly with good reproducibility of the values. The interference by other elements is negligible and the detector limits are better than any other instrumental technique. At the same time, the standardisation and operation of the equipment is also simple and as such thousands of samples could be analysed accurately within a short time. In view of its versatility and sophistication, the equipment is very expensive and very few laboratories have acquired it and maintained it.

The absorption of light energy by the free atoms in the ground state is measured in this instrument and this then correlated with the concentration. When a salt solution is introduced into a flame, the molecules of the salt dissociate into neutral atoms (fundamental, ground or unexcited state), excited neutral atoms and ions. In this instrument, the number of fundamental state atoms in the flame is measured and these atoms are far more in excess of the number of excited atoms. (In flame photometry, the radiation from the excited atoms are measured).

When a light radiation of appropriate wavelength is passed through a cloud of atoms in the flame, the neutral fundamental state atoms absorb the radiations resulting in these atoms reaching the excited state. The absorption follows Lambert - Beer's law and holds good only for a specific wavelength (In absorption spectroscopy, it is molecular absorption).

The excited atoms emit radiations when they return to the ground state since the wavelength of the radiations is same as the incident radiation (absorbed radiation). Emission will interfere with the absorbance measurement. In Atomic Absorption Spectrophotometer, this is overcome by using a light source of particular pulse frequency and matching it with an AC amplifier which responds to the signals of the particular frequency only. Thus the emitted radiation which is a continuous emission will not be measured by the photometric system.

An Atomic Absorption Spectrophotometer consists of the following components :

i. A light source emitting a sharp line spectrum of the element to be detected.

ii. A mechanism for producing an atomic vapour of the sample.

iii. A wavelength selector to separate the resonance line requirement.

iv. A detector - amplifier and read out system.

Light source

The light source emits resonance radiation of the element and is sufficiently intense and stable to permit reliable measurements. Also the background emission is maintained at low level. Hollow cathode lamps with the element to be analysed for as cathode and fitted with an inert gas at low pressure is used to give a characteristic spectrum of the colliding material. This spectra contains the resonance wavelength radiation along with other radiations.

Atomisation system

This consists of a nebulizer to provide a free fine fog of droplets from the test solution and the burner. The unit is the same as in the flame photometer except that the burner has a long narrow slit on head just like a fish tail burner, so that the flame provides a long absorption path for the incident radiation. Various fuel - oxidant systems like acetylene - air or oxygen, nitrous oxide - hydrogen - air or oxygen, propane - air or oxygen have been used for the individual element analysis. The fuel - oxidant systems suitable for various elements are as given in next page.

Monochromator

This isolates the resonance wavelengths from other

radiations before measurement of the intensity. A diffraction grating or prism can be used and filters are not used in this equipment.

Fuel-oxidant systems	Elements that can be analysed
Acetylene - air	Sn, Mo, Cr, Cu, Fe, K and Mg; Suitable absorbance between 0.1 to 0.2
Acetylene - nitrous oxide	Mg, Ca, Sb, Ba, Sn, Cr, B, Mo.
Hydrogen - air	As, Se, Te, Sn.
Nitrogen - hydrogen entrained air	As, Se, Cd, Hg, Sn, Te, Ca, Pb, Zn
Argon - hydrogen entrained air	Has a slightly higher transparency in the far off UV region and is slightly less prone to interference.
Propane - nitrous oxide	Mg and Ca in blood.

Detector Unit

This consists of photosensitive cell (photomultiplier), an amplifier and a sensitive galvanometer. The analysis of nonmetals are not done easily in this instrument as their resonance lines occur in the vacuum UV region. Hence indirect methods have been developed based on their effects on anions on the atomic absorption spectrometer.

Generally the hollow cathode lamp is set up first and then the instrument is adjusted to the desired wavelength to get the maximum reading on the indicator unit. With the flame burning, the gains control is then adjusted to get a reading of 100 in the transmittance. Then the final adjustment is made with the fine gains control. The stability of the instrument should be checked and if necessary adjusted with the damping control. With the flame burning, and after closing the filter, the meter is adjusted to zero reading and the filter is reset to open. By placing a shutter (piece of cardboard) between the lamp and the flame, check the meter to see whether it reads zero, indicating that the signal due to the noise caused by emission from the flame is negligible. If the zero value is not obtained then the instrument has to be adjusted to zero with the tracking control. Then the shutter switch should be put in absorbance and the meter set to zero with the fine gains control. After checking these, the solution may be sprayed into the flame and reading noted.

INFRA RED ABSORPTION SPECTROSCOPY

Light can be divided into ultraviolet, visible and infrared regions. The last region can be further divided into four regions: (photographic region-visible to 1.2 μ; very near infrared (overtone region) - visible to 2.5 μ; near infrared (vibration rotation region) - 2.5 to 25 μ and far infrared (rotation region) 25μ to 300-400 μ}. In the infrared absorption spectroscopy methods, the near infrared region is made use of. The usual optical materials, glass or quartz, absorb strongly in the infrared region. Consequently the apparatus used is different from the ones used for visible and UV regions. The main difficulty in the infrared spectrophotometry is the need for using a narrow cell (thickness of 1 mm or less). Absorption or extinction coefficient cannot be determined precisely owing to the difficulty of making accurate measurements of sample thickness and also the necessity to use wide slit width, since the energy available in the useful wavelength range is relatively small. Even a very slight shift in wavelength setting would cause a considerable change in the apparent value of the absorbing index or specific extinction coefficient.

Hence an empirical procedure known as base line method is adopted in the estimation of a concentration of single component in a multicomponent system. The substances already present in the sample serves as standards in determining the compounds of interest. In this the spectra for all compounds to be determined in the mixture is obtained and then the appropriate absorption peaks of each one of the compounds is selected such that there is minimum interference by the peaks of other compounds. Now most of the infrared spectrophotometers in use are of double beam reading type and are expensive and are used mostly in the analysis of organic compounds.

CHAPTER - 8

PRESENTATION OF THE ANALYTICAL DATA

Every analyst analyses large number of samples daily and throughout the determinations, his main concern is to get the most reliable values. As soon as he has achieved the same, he just proceeds to present them in his own way, without realising its impact on the reader. In any analytical work, proper presentation of the data is as important as carrying out the estimations accurately. Many a time, this aspect is overlooked by many analysts and improper presentation of the data leads to wrong interpretation of the values in number of cases.

The analyst commonly uses two terms, precision and accuracy interchangeably to describe the reliability of the results without understanding the basic difference in their meaning. Actually both the terms do not mean the same thing. Precision is defined as the concordance of a series of measurements of the same quantity. Mean deviation or relative mean deviations is a measure of precision. It is also defined as the degree of reproducibility of a series of measurements of the same property, that is, a result is said to be precise if number of parallel determinations agree within limits. For example in the following data it can be seen that the per cent variation in the case of analyst "A"

A	B
34.48	34.58
34.47	34.42
34.49	34.62

is much less than that of analyst "B" and hence the results of analyst "A" can be said to be more precise than that of analyst "B". But at the same time, the values obtained by "A" need not be accurate, that is close to the true value or the most probable value. Precision always accompanies accuracy but a high degree of precision does not imply high accuracy as could be seen from the calculations given below :

Relative mean deviation

Weight of a compound	Deviation
48.32	0.04
48.36	0.08
48.23	0.05
48.11	0.17
48.38	0.10
241.40/5	0.44/5
Mean 48.28	0.09

$$\text{Relative mean deviation} = \frac{\text{Mean deviation}}{\text{Mean}} \times 100 = \frac{0.09 \times 100}{48.28}$$

$$= 0.19\%$$

The analytical values furnished by two analysts are given below; The substance is known to contain 49.06 grams of the constituent.

Analyst A		Analyst B	
Values	Deviation	Values	Deviation
49.01	0.09	49.40	0.02
49.21	0.11	49.44	0.02
49.08	0.02	49.42	0.00
Mean 49.10	0.22/3	49.42	0.04/3
	0.073		0.013

$$\text{Relative mean deviation} = \frac{0.22 \times 100}{3 \times 49.10} \qquad \frac{0.04 \times 100}{3 \times 49.42}$$

$$= 0.15\% \qquad = 0.05\%$$

$$\text{Relative mean error} = \frac{49.10 - 49.08}{49.08} \qquad \frac{49.42 - 49.06}{49.06}$$

$$= 0.08 \qquad = 0.73$$

	Relative mean error (Accuracy)	Relative mean deviation (Precision)
A	0.08	0.15
B	0.73	0.15

From the above, it can be seen that the values obtained by the Analyst A is accurate and precise while those obtained by Analyst B unusually precise but less accurate.

As could be seen from the values given above, in any analysis involving biological materials, invariably the analytical values obtained from parallel determinations do not agree and there will be some deviation. There are occasions when one of the values appears to be far out of the way from those of other values leading to the possibility of rejecting the same by visual judgement. This may lead to bias in the estimations and may vitiate the final outcome. Hence certain procedures have been suggested to include or reject the particular suspected value for calculating the means.

Procedure for rejection of data

For rejecting or accepting any piece of data in a set of values, first a rejection quotient (Q) is calculated for a small number of observations. For this the data is arranged first in decreasing order of the values. Then the difference between the suspected value and its nearest neighbour is divided by the range (difference between the highest number and the lowest). The ratio is compared with the tabulated values of Q. If the calculated Q value is equal to or greater than the tabulated values for a given number of observations, then the value in question may be rejected.

Confidence level : Rejection quotient (Q) for 90%.

No. of observations	Q	No. of observations	Q
3	0.94	8	0.47
4	0.76	9	0.44
5	0.64	10	0.41
6	0.56	0	0.00
7	0.51		

Example

An analyst reports the following four values in respect of a particular analysis; 102, 107, 108, 115. The last value 115 appears to be out of the general trend.

Its nearest neighbour is 108 and the difference is 7. The range is 115-102 = 13. Rejection quotient 7/13 = 0.54. From the table given above the "Q" value for four observations is 0.76. As the calculated value is less than the Q value given in the table, the suspected value cannot be rejected.

In another method that is used to overcome the bias, the

doubtful value is omitted and then the mean and the mean deviations of the other retained values are determined. If the deviation of the suspected value from the mean is atleast four (4) times the mean deviation of the other retained value (>4 x mean deviation), then the doubtful value may be rejected. For example ; the following values were obtained in a test. 46.34, 46.47, 46.64, 46.76, 46.71, 46.53, 46.60, 46.60, 46.71 and 46.62. Suppose the suspected value is 46.34 then the mean omitting the suspected value is 46.627 and the deviation of the suspected value from the mean is 0.287 (46.627 - 46.34 = 0.287). Mean deviation is 0.07. Then four times the mean deviations is 4 x 0.07 = 0.28. As the calculated value is more than 0.28, then the suspected value has to be rejected and only the other values presented while presenting the data.

Even in presenting the valid data, certain procedures have been proposed to provide meaningful data with better clarity and at the same time avoiding unnecessary details as well as misleading information.

Significant figures : A significant figure denotes the quantity in the place in which it stands (units, tens, hundreds etc.). The significant figures in a number are the digits necessary to express the results of a measurement to the level of precision with which it is made. For example a weight of a weighing bottle in a physical balance may be 14.3 gm and when weighed in an analytical balance its weight may be 14.3263. In the former case, there are three significant numbers and in the latter case six significant numbers.

The digit zero "0" is a significant figure except when it is the figure before the decimal. For example in 0358 and 358, there are only three significant figures as the zero in the first case has no significance. But in the figure 0.000358, the three zeros after the decimal indicates that magnitude of the significant number and could be written as 358 x 10^{-6}. Excluding the exponential factor, (10^{-6}) the result has only three significant numbers. Exponential factor (10^{-6}) serves only to locate the decimal points and does not fit into the definition of significant numbers. Hence the significant figures in a number includes all of the certain figures and also the first uncertain figure or doubtful figure of the number. For eg. in the figure 0.000358, the certain numbers are 0.00035 and the uncertain number is 8. As such with a more accurate equipment it can be anywhere between 0.000358 and 0.000359. Unless the uncertainty of the doubtful figure is stated, it is generally assumed to be ± 1 of the last digit.

In practice, it is very common to find analysts attempting to give values to as many significant numbers as possible to give an impression of having done the analysis with the greatest possible accuracy. However in most of the cases, certain figures do not have any meaning at all. The value that is given indicates only the level of the sensitivity of the instrument used and nothing more. There is always a possibility of getting a still more accurate value of the same parameter had the analyst used an instrument of greater precision and sophistication than the earlier ones.

In the example given earlier, the weight recorded by the physical balance indicates that the weight of the material lies between 14.2 and 14.4 gm. While that by the analytical balance the weight is indicated to lie between 14.3262 and 14.3264 gm, if the sensitivity is ±1. If the sensitivity is ±4, then the weight is between 14.3259 and 14.3267 gm. Aiming at this degree of accuracy is needed in certain cases, and unwarranted and meaningless in others. Hence to present the data properly and meaningfully and also to avoid misleading conclusions, the following guidelines in the use of significant figures and also for computing them have been proposed;

a. First the nature of graduations or accuracy of the measurements of the equipments should be noted. For example the titer value using a burette graduated in 0.1 ml level has to be written as 16.5 ml and not as 16.50 ml. In the former case it will be assumed that the titre value is between 16.4 and 16.6 ml with a mean of 16.5. But in the latter case when it is written as 16.50, it will mean that the titre value is between 16.49 and 16.51 ml indicating the accuracy of the burette as 0.01 ml which is not true. Hence in this case the titre value has to be written only as 16.5 and not as 16.50.

At the same time if a microburette with graduations of 0.01 ml was used in the estimation and the titre value was exactly 16.5 ml, then it must be written as 16.50 and not as 16.5 ml indicating that the accuracy of the burette was ± 0.01 ml and hence the correct value can be anything between 16.49 and 16.51 ml.

b. It is a common experience to find values reported with four or more significant figures after the decimal and two or three before the decimal. In most cases aiming at such accuracy has no relevance when one considers the other possible experimental errors. For example the weight of a precipitate may be reported as 14.3728 gm. This may mean that the analyst had at his disposal an analytical balance capable of weighing upto 0.1 mg and has taken care to weigh accurately to the fourth decimal place of a gram. But this may not have any relevance in number of instances as can be seen below; Even if he had recorded the same value (14.3728) as 14.3800 or 14.38, the difference between the two values would be 0.0072 and in percentage it will be

$$\frac{72 \times 100}{143728} = 0.005 \text{ per cent}$$

which is far below the possible analytical errors in any estimation. In case he had recorded just upto even one significant figure after the decimal, say as 14.4 gm instead of 14.37 gm, the error would have been only 3 parts out of 1437 parts which is less than even 1 per cent error, and hence weighing or aiming at this amount of accuracy has no meaning. At the same time if the material absorbed is in milligrams only as in the case of ethylene glycol vapour on the surface of soil in the measurement of surface area, the material has to be weighed to the best possible accuracy as only few milligrams of the solvent vapour will be absorbed on the surfaces. In such cases, use of sophisticated instruments and reporting to the last possible decimal is fully justified. Hence while recording the values one has to consider the degree of accuracy needed for clarity and attempt only to that level. This is particularly so when dealing with the biological systems. No useful purpose will be served in recording data at greater accuracy level than that is needed as it may just indicate more digits in the figure and in cases where means, deviations etc., have to be worked out, may consume more of computational time which is unnecessary. Whenever cost per analysis is worked out, it will add to the already high costs.

Rounding of figures : Just as in the presentation of data upto significant level, in computations where numbers have to be rounded off at the last but one significant number, the following rules may be followed:

a. If the last figure is less than 5, then the last but one figure is retained as such. For example in 4.3243, if this has to be rounded of to three decimals, then it will be 4.324 and not 4.325.

b. In case the last figure is more than 5 then the last but one figure is raised to the next number. For example in 4.3246, it may be rounded to the third decimal and in that it will be 4.325.

c. When the last figure is 5 and there are no figures later or there is only zero beyond 5, then the figure in the last place to be retained is increased by one if it is odd. In case it is even, then it is not changed at all. For example in the figure 1.235 or 1.2350 it will be rounded of to 1.24 if only two significant numbers after the decimal are needed. In case the number is 1.2450 then it will be rounded of to 1.24 and not to 1.25 since the last figure to be retained is even.

d. When the figure other than zero is beyond 5, then the last figure to be retained is increased by one irrespective of whether it is even or odd. For example in the case of 1.2451, it will become 1.25 if only two decimals have to be retained.

e. In addition and subtraction, computations of significant figures to each number and in that the sum or difference is written only to the place in which an uncertain figure occurs farthermost to the left relative to the decimal point in any number. For example

65.43		65.43
01.245	becomes	01.24
00.4652		00.47
67.1402		67.14

f. In multiplication and division also, in reporting the final result, only as many significant figures as are contained in the term having the largest relative uncertainty are retained and used for calculations. For example in the multiplication 0.156 x 84.25 x 1.7854, the same is reduced to 0.156 x 84.2 x 1.79, assuming the last digit in each number to be doubtful by one unit, then the precision of the original numbers are 1/156. 1/8425, 1/ 17854. The first number (1/156) has the largest uncertainty compared to others and is the least precise of all the three of values. Hence it determines the number of figures to be retained in the other terms and the final value after multiplication or division as the case may be.

g. In writing the final result of multiplication, the agreement among the samples must be considered. For example the average of 45.02, 45.34 and 45.18 in a sample should be reported as 45.2 and not 45.10 because the second figure after the decimal is uncertain in this analysis. However the average of 45.120, 45.180 and 45.160 should be reported as 45.15 and not as 45.153 as the third figure after the decimal is uncertain.

SECTION - B

METHODS OF ANALYSIS

CHAPTER – 9

ANALYSIS OF WATER

Every naturally occurring water contains varying quantities of dissolved solids and gases and in a few cases, some amount of suspended organic and/or inorganic materials. The nature, amount and relative proportions of these influence the quality of waters. The term "quality " and its specifications depend upon the purpose for which the water is intended to be used. A sample of water which is classified as "unfit" for certain industrial processes may be classified as "fit" or "good" quality for certain other purposes. Hence the test to be carried out to assess the quality of water depends on the purpose for which it is likely to be used and also on the nature of ions present and the proportion in which they occur with respect to each other. In the case of irrigation waters, the nature and concentration of various ions, particularly the proportion of divalent ions to the monovalent cations, are of great importance in assessing the quality as these ions affect the crop growth differently. The common cations that are present are calcium, magnesium, potassium and sodium and the anions commonly found in detectable levels are chloride, sulphate, carbonate and bicarbonate. In addition to the above, boron and nitrate-nitrogen are also found in a few cases and are estimated if suspected to be above toxic levels as many crops are susceptible to the presence of these ions at toxic level. Normally the waters are analysed for the above cations and anions and based on their contents, their quality is fixed.

The most important step in the analysis of water samples is the collection of representative water sample from the well or tank or pond. In number of instances the reported quality of water is found to be in error inspite of the fact that the analysis has been done in a reputed laboratory by an expert analyst. This is due to nonobservance of certain basic points in collection of the samples and also in storing them before analysis.

COLLECTION OF WATER SAMPLES

Water samples should be collected in either a glass or polythene stoppered bottle preferably in a transparent one. It should be thoroughly cleaned before use and should be rinsed 3 to 4 times with the water which is to be sampled. When new (fresh, unused) bottles or containers are to be used, the cleaning and washing has to be more rigourous and it is preferable to keep them filled with water for a day or two, then clean them again before using. Care has to be exercised to clean the crevices inside the bottle for any possible sediment or precipitate. Use of plastic corks should be advocated wherever possible. Porous corks and rubber corks should be avoided to the possible extent. In case they have to be used due to nonavailability of the desired ones, then they should be covered with one or two sheets of clean, uncontaminated plastic paper so that the water sample does not come in contact with the cork. Special care has to be taken to see that these corks are clean as in number of standard analytical laboratories, the number of samples discarded due to this reason are more than the samples analysed. Normally one liter of water sample would be sufficient for all analysis.

Sampling of water

If the water sample has to be collected from tubewell, hand pump etc., they must be run for atleast 15 minutes so that the sediments, precipitates already formed either on the surface of the well or in the pipelines due to drying of materials, are washed away and are prevented from contaminating the sample. Wherever possible, it may be preferable to collect larger quantity of water in a clean plastic bucket and then fill up the bottle.

If the water has to be collected from an open well, a bottle or bucket with the lid closed, may be lowered to at least 2 to 3 meters (5 to 10 feet) below the surface without opening the lid. Containers made up of light materials will float on the surface and will not get immersed due to its low density and light weight. In such cases, a heavier material like stone may be tied to the outer bottom of the container and then the container lowered. The water sample should be collected preferably from the centre of the well and never near the walls of the well.

If the source of irrigation water is an open tank, canal or river, the sample should be drawn from a spot in the midstream, away from the sides with a clean bucket previously rinsed with the same water. The sample thus drawn should be immediately transferred to a clean bottle and stoppered tightly. In many cases, the samples are transferred or collected in a container observing all precautions but left in the open without putting the lid or stopper. In such cases, the composition is likely to be affected either due to the contamination from outside or due to the evaporation of water which would concentrate and thus result in an increase in the concentration of ions. So the containers should be stoppered immediately with a clean stopper after the collection of the sample. Then a proper label should be prepared indicating the source, name of the owner, crops to be raised with water (purpose for which the water is going to be used), nature of soil and any other information like depth, problems encountered if any, in earlier seasons (or uses) etc. The water sample should be analysed immediately and in case a few days of delay in taking up the analysis is inevitable, then 2 or 3 drops of toluene may be added to prevent bacterial activity. If there is any necessity to filter, this may be done using Whatman No. 1 filter paper.

ANALYSIS OF WATERS

The water samples are analysed for the following : pH, total soluble salts (electrical conductivity), calcium, magnesium, sodium, potassium, carbonate, bicarbonate, chloride, sulphate, boron, nitrate-nitrogen. The methods of analysis detailed below are simple and can be carried out in any analytical laboratory.

1. pH

A water sample (approximately 100 ml) may be taken in a clean beaker and the pH is determined with a pH meter. While determining the pH of the sample, care should be taken to see that the electrodes dip well into the sample. Also before starting with the samples, the pH meter should be checked with atleast three buffer solutions, two on the acid range and one on the alkaline range. If large number of samples are being estimated, then the buffers should be introduced after each batch of 10 or 15 samples and checked. If any deviations are observed then the instrument should be checked and then only further analysis taken up.

2. TOTAL SOLUBLE SOLIDS

i. Gravimetric method

Transfer a known aliquot (100 ml) of the water sample to a previously weighed potash basin and evaporate it to dryness on a water bath. In this analysis, use of silica basins or heavy weight containers should be avoided as it will be difficult to observe small differences in weights caused by the solids in water. Potash basins or other containers of light weight that do not react with the water may be preferred. Care should be taken to see that the bottom of the basin does not come into contact with the water in the waterbath. Also greater the quantity of water sample taken, greater will be the accuracy. The potash basin with the salt of the water (dried water sample, after evaporation) should be kept in an air oven for one hour, then transferred to a desiccator and cooled. After it has attained the room temperature, find out the weight of the basin plus the dried water sample. From these weights, calculate the total solids present in the water sample.

Calculations

Aliquot of water sample taken = V ml

Weight of empty potash basin = W_1 gm

Weight of potash basin + water sample (after evaporation) = W_2 gm

Weight of total solids in the water sample = $(W_2 - W_1)$ gm

Total solids in ppm $= \dfrac{(W_2 - W_1) \times 10^6}{V}$ ppm

ii. Electrical conductivity method

The amount of total solids may be assessed from the electrical conductivity of the water sample using a solu-bridge or any conductivity meter. Clean the cell of the conductivity meter or solu-bridge with distilled water and fill the same with water sample after rinsing the cell atleast twice with the test sample. Then the solu-bridge should be adjusted to read at the room temperature. Find out the electrical conductivity and express it in millimhos/cm or if it is very low, in micromhos/cm (one thousandth of millimhos/cm). As in the pH determinations, introduce check sample of known conductivity before and after every 10 or 15 samples and check the accuracy of the instrument. The electrical conductivity is approximately related to the total soluble salt content of irrigation waters.

3. CARBONATES AND BICARBONATES

Principle

The carbonate and bicarbonate ions are present as salts of strong bases and weak bases and are alkaline in reaction. These ions are estimated by titration with a mineral acid using phenolphthalein and methyl orange as indicators. In the first part of the titration using phenolphthalein as indicator, the carbonates are converted to bicarbonate. In the second part of the titration using methyl orange as indicator, all the bicarbonates that have been formed is neutralised by the acid.

$$2Na_2CO_3 + H_2SO_4 \dashrightarrow NaHCO_3 + Na_2SO_4$$

$$2NaHCO_3 + H_2SO_4 \dashrightarrow Na_2SO_4 + 2H_2O + 2CO_2$$

In this one molecule of sulphuric acid reacts with two molecules of carbonate in the presence of phenolphthalein. Similarly all the bicarbonates originally present and formed from the conversion of carbonate to bicarbonate reacts with sulphuric acid to produce sodium sulphate.

Reagents

Sulphuric acid (0.1 N), phenolphthalein indicator, methyl orange indicator

Procedure

Pipette out 50 ml of the water sample into a 250 ml conical flask. Add 2 drops of phenolphthalein indicator and titrate it against 0.1 N sulphuric acid till the disappearance of the pink colour. Record the volume of standard acid consumed (X ml).

To the colourless solution from the above titration or to the original sample of water if no pink colour was observed with phenolphthalein, add 1 to 2 drops of methyl orange indicator and continue the titration with constant stirring till the appearance of yellow colour. Note the volume of acid consumed in this titration with methyl orange (Y ml). From these values calculate the amount of carbonate and bicarbonate present in the water sample.

Calculations

Volume of water taken = V ml

Volume of 0.1 N sulphuric acid used with phenolphthalein as indicator (CO_3 to HCO_3) = X ml

Volume of 0.1 N sulphuric acid used with methyl orange as indicator = Y ml

Volume of 0.1 N sulphuric acid required to neutralise bicarbonate alone = (Y – X) ml

Volume of 0.1 N sulphuric acid required to neutralise the carbonate alone = (2 x X) ml

i. Carbonate

1 ml of 0.1 N sulphuric acid = 0.003 gm CO_3

(2 x X) ml of 0.1 N sulphuric acid = 0.003 x(2X) gm CO_3

This is present in V ml of sample

Hence in 10^6 ml of water, carbonate present $= 0.003 \times 2X \times \frac{10^6}{V}$ ppm

Milliequivalents of CO_3 present/liter

$$= \frac{0.003 \times 2X}{V} \times 1000 \times \frac{1000}{30}$$

$$= \frac{2\ X \times 100}{V}$$

If the normality of the acid used is different then

$$\text{Milliequivalents of } CO_3\text{/litre} = \frac{2X \times 1000 \times \text{Normality of acid}}{\text{Volume of water taken}}$$

ii. Bicarbonate

1 ml of 0.1 N sulphuric acid = 0.0061 gm of bicarbonate

(Y-X) ml of 0.1.N sulphuric acid = 0.0061 x (Y-X) gm of HCO_3

This is present in V ml of water.

Hence in 10^6 ml of water, the bicarbonate present $= \frac{0.0061 \times (Y - X) \times 10^6}{V}$ ppm

Milliequivalents of bicarbonate /litre

$$= \frac{0.0061 \times (Y - X) \times 1000}{V} \times \frac{1000}{61}$$

$$= \frac{(Y - X) \times 100}{V}$$

If normality of acid used is different then bicarbonate in milliequivalents /litre

$$= \frac{(Y - X) \times 1000 \times \text{Normality of acid}}{\text{Volume of water taken}}$$

4. CHLORIDES

Principle

Chlorides are invariably present in all natural waters and contribute to its salinity. The chloride content is estimated by Mohr's method wherein the chloride is precipitated as silver chloride using silver nitrate. After all the chlorides have been precipitated, the excess silver nitrate combines with potassium chromate that has been added as indicator to the water to form a reddish brown coloured precipitate.

$NaCl + AgNO_3$ ------> $AgCl + NaNO_3$ (white precipitate)

$K_2CrO_4 + 2\ AgNO_3$ ----> $Ag_2CrO_4 + 2\ KNO_3$
(reddish brown precipitate)

Reagents

Silver nitrate (0.02 N), potassium chromate indicator

Procedure

Pipette out 50 ml of water sample (or even the same solution after the carbonate and bicarbonate titration may be used in case the water sample available for study is limited) into a clean white porcelain basin. Titrate it against 0.02 N silver nitrate solution after adding potassium chromate indicator to the water. Titration should be done with constant stirring till a brick red colour is obtained. Record the volume of standard silver nitrate solution used. With this value calculate the amount of chloride present in the waters.

Calculations

Volume of water taken = V ml

Volume of 0.02 N silver nitrate used = X ml

1 ml of 0.02 N silver nitrate = 0.00071 gm of chloride

X ml of 0.02 N silver nitrate = 0.00071 x X gm of chloride

This is present in V ml of water sample

Hence in 10^6 ml of water the chloride content

$$= \frac{0.00071 \times X \times 10^6}{V} \text{ ppm}$$

In terms of milliequivalents /litre chloride content

$$= \frac{0.00071 \times X \times 1000}{V} \times \frac{1000}{35.5}$$

In case silver nitrate of different normality has been used

$$= \frac{X \times \text{normality of } AgNO_3 \times 1000}{\text{Volume of water in ml}}$$

If the quantity of chloride present in one litre is needed :

1 ml of 0.02 N silver nitrate = 0.00071 gm chloride

X ml of 0.02 N silver nitrate = 0.00071 x X gm of chloride

This is present in "V" ml of water sample

Hence in 1000 ml of water sample the amount of chloride present

$$= \frac{X \times 0.00071 \times 1000}{V} = \frac{0.71\,X}{V} \text{ gm}$$

5. SULPHATE

i. Gravimetric method

Principle

Sulphate is present in most of the water samples either as sodium sulphate or calcium sulphate or magnesium sulphate. Sulphates in water can be estimated gravimetrically or volumetrically. In the gravimetric method, the sulphate is precipitated as barium sulphate by adding barium chloride to the water sample in acidic solution. The precipitated barium sulphate is filtered, washed, ignited and weighed.

Reagents

Concentrated hydrochloric acid, barium chloride (10%), methyl orange indicator.

Procedure

Take 100 ml of the water sample and add a few drops of methyl orange and 1 ml of concentrated hydrochloric acid. Heat to boiling and add an excess of barium chloride drop by drop with constant stirring. Allow it to stand on a water bath until the volume is reduced to 50 ml. After cooling, the precipitate of barium sulphate is filtered through Whatman filter paper No.3 (ashless) and washed with distilled water until it is free of chloride. Then fold the filter paper carefully, keep it in a previously weighed silica crucible, ignite it slowly in a low flame initially and then at high temperature. The temperature should not exceed 900°C but it is preferable to oxidise it at temperature around 600°C. When most of the paper has been oxidised, the temperature can be raised to remove the last traces of carbon. Then cool the silica crucible with the precipitate and find out the weight of barium sulphate precipitate.

Calculations

Volume of water taken = V ml

Weight of empty silica crucible = W gm

Weight of silica crucible + precipitate = W_2 gm

Weight of barium sulphate precipitate = $(W_2 - W_1)$ = W gm

This is found in V ml of water. Hence

Amount of barium sulphate in 1000 ml = $\frac{W \times 1000}{V}$ gm

Weight of sulphate in 1000 ml = $\frac{W \times 1000 \times 0.412}{V}$ gm

(Conversion factor for barium sulphate to sulphate = 0.412)

Amount of barium sulphate in 10^6 ml = $\frac{W \times 10^6}{V}$ gm

Amount of sulphate in 10^6 ml = $\frac{W \times 10^6 \times 0.412}{V}$ gm

ii. Volumetric method (Jackson, 1973)

Principle : In this also, the sulphate is precipitated as barium sulphate using barium chloride and the amount of sulphate is estimated by the EDTA titration method.

Reagents

Barium chloride (0.02 N), magnesium chloride (0.02 N), EDTA (0.02 N), buffer solution, calcium chloride (0.02 N), erichrome black indicator, hydrochloric acid (0.02 N)

Procedure

Pipette out an exact quantity of water sample into a 250 ml beaker and add a measured volume of standard hydrochloric acid equivalent to the total alkalinity (amount of HCO_3^- plus CO_3^{2-}) and boil it to destroy carbonate and also drive away carbon dioxide. Then add a known volume (25 ml) of 0.02 N barium chloride and boil it for a few seconds. After cooling add 10 ml of the buffer and 5 drops of erichrome black T indicator and titrate it against EDTA solution to the first end point. As accurate end point cannot be detected at this stage even if the versenate solution contains some magnesium, add a known quantity of standard (0.02 N) magnesium chloride and titrate it to the end point as in the case of calcium and magnesium.

Calculations

$$\text{Milliequivalents of sulphate} = B + B_a + M_q - T$$

Where

B = Blank. Meq of Ca^{2+} and Mg^{2+} in original solution

B_a = Milliequivalents of barium chloride added

M_q = Milliequivalents of magnesium chloride added

T = Milliequivalents of versenate used in the total titration of the sample after adding barium chloride and magnesium chloride.

If the amount is required in ppm, then

$$\text{ppm} = \frac{\text{Milliequivalents}}{\text{Equivalent weight}}$$

6. BORON

Principle

Normally waters intended for irrigation are not tested for boron or nitrate-nitrogen unless they are suspected to be present in appreciable quantities. Some waters particularly those from entrapped underground sources or springs may contain excessive amounts of boron. Of the number of organic compounds that form coloured complexes with boron as borate, Curcumin $CH_3OC_6H_3$ (OH)CH:CH= $CO_3.AH_3$, a dye extracted from turmeric is preferred as it gives a sharp colour differentiation. As coloured lakes or compounds are formed with barium, iron, aluminium or magnesium in alkaline conditions, for boron estimation an acid medium is preferred. If nitrates are present in quantities more than 90 ppm, they can be eliminated by evaporating the water to dryness after adding 10 ml of saturated solution of pure calcium hydroxide followed by ignition and extraction with 0.05 N hydrochloric acid. In this estimation, glasswares made of hard and resistant glass (borosilicates) should not be used. Only specially made boron free glasswares like ordinary soda-glass or silica or polythene wares should be used.

Reagents

Hydrochloric acid (0.05 N), ethyl alcohol (95% boron free), Curcumin-oxalic acid solution, standard boron solution (100 ppm).

Equipments

Spectrophotometer

Procedure

Take a known quantity of water sample containing 0.2 to 5 μg of boron in 250 ml beaker and add 0.05 N hydrochloric acid till it is acidic. Then add 4 ml of curcumin - oxalic acid reagent and evaporate to dryness at 55 - 60°C on a hot plate or water bath. Keep it for another 15 minutes at that temperature. After cooling, mix the residue thoroughly with 25 ml of 95% ethyl alcohol to extract the rose cyamine coloured complex and filter it through Whatman No. 1 filter paper directly into a colorimetric tube. Measure the intensity of the solution using spectrophotometer and at 540 mμ within 2 hours. Using a standard curve, find out the amount present in the water sample.

Calculations

$$\text{Boron (ppm)} = \frac{\mu\text{g of boron read on the standard curve}}{\text{mg of water sample taken}}$$

7. NITRATE - NITROGEN

Principle

Normally waters do not contain appreciable amounts of nitrate - nitrogen. However rain water received from certain areas located near areas contaminated with nitrate - nitrogen may contain appreciable amounts of nitrate - nitrogen. Beneficial effects of nitrate - nitrogen have been reported in brackish waters. It is estimated by the phenol disulphonic acid method. A nitrate in dry condition, in the presence of concentrated sulphuric acid in cold, causes nitration of phenol disulphonic acid, giving intense yellow colour in alkaline medium which can be measured colorimetrically at 420 mμ. Using a standard curve, the amount present can be found out and from that the amount present in either ppm or in per cent can be calculated.

Reagents

Calcium sulphate or calcium oxide, ammonia (1:1), phenol disulphonic acid.

Procedure

Take 50 ml of water sample in a porcelain dish or beaker and evaporate the same to dryness on a water bath and add 3 ml of phenol disulphonic acid reagent and allow it to react for 10 minutes. To this add 15 ml of distilled water and stir well with a glass rod. Transfer the contents to a 100 ml volumetric flask and then add 1:1 ammonia slowly with mixing till the solution is alkaline in reaction as indicated by the yellow colour due to the presence of nitrates. Then add 2 ml of ammonia and make upto volume with distilled water, shake it and read the intensity of yellow colour in a photoelectric colorimeter using a 420 mμ filter. From the standard curve already prepared, find out the amount present in the extract and calculate the amount present in a given volume.

Calculations

$$\text{Nitrate - nitrogen (ppm)} = \frac{\mu\text{g of } NO_3 - \text{N in solution}}{\text{Volume of water sample taken}}$$

8. CALCIUM AND MAGNESIUM

Principle

Calcium and magnesium when reacted with EDTA get complexed in the following order; calcium and then magnesium. Both are estimated by the erichrome black T indicator. The formation of magnesium complex is optimum at pH 10 in the presence of ammonium chloride plus ammonium hydroxide. Erichrome black T indicator gives a wine red coloured complex with magnesium but changes into sky blue when all the magnesium ions have been removed by the versenate. Calcium is estimated first using murexide indicator which forms an orange red complex of calcium ion at pH 12 and this gets converted into a red-violet colour when all the calcium ions have been completely complexed by the versenate.

Reagents

EDTA (0.01 N), ammonium chloride - ammonium hydroxide buffer pH 10, erichrome black T (Solochrome black) indicator, murexide (ammonium purpurate) indicator, sodium hydroxide (approximately 4 N), standard calcium chloride (0.01 N).

Procedure

i. *Calcium plus magnesium*: Take 25 ml of water sample in a porcelain dish and dilute it with 25 ml of distilled water. Add one ml of ammonium chloride - ammonium hydroxide buffer and 3 to 4 drops of erichrome black T indicator and titrate against 0.01 N EDTA (versenate) solution, till the colour changes from wine red to blue or bluish green. At the end, no tinge of red colour should be seen. Note the volume of EDTA used.

ii. *Calcium*: Take the same aliquot of the sample (25 ml) in another porcelain dish and dilute it with 25 ml of distilled water. Add 5 drops of approximately 4 N sodium hydroxide and approximately 50 mg of purpurate indicator. Titrate the solution with standard EDTA till the orange red colour changes to purple. As the colour change is not instantaneous, add EDTA drop by drop and with constant stirring after each drop. For comparison prepare a blank by adding sodium hydroxide indicator and a drop or two of EDTA to distilled water and keep it for comparison.

Calculations

Volume of water sample = 25 ml

Volume of 0.01 N EDTA used for (Ca+ Mg) = A ml

Volume of 0.01 N EDTA used for calcium alone = B ml

Volume of 0.01 N EDTA used for magnesium alone = (A–B) ml

Calcium

1 ml of 0.01 N EDTA = 0.0002 gm of Ca

B ml of 0.01 N EDTA = 0.0002 x B gm of Ca

This is present in 25 ml of water. Hence

In 1000 ml of water = $\frac{0.0002 \times B \times 1000}{25}$ gm of Ca/litre

In 10^6 ml of water = $\frac{0.0002 \times B \times 10^6}{25}$ gm of Ca

In milliequivalents per litre = $\frac{0.0002 \text{ x } B \text{ x } 1000 \times 1000}{25 \times 20}$

$= \frac{B \times 10}{25}$

Magnesium

1 ml of 0.01 N EDTA = 0.00024 gm of Mg

(A-B) ml of 0.01 N EDTA = 0.00024 x (A-B) gm of Mg.

This is present in 25 ml of water. Hence

1000 ml of water contains = $\frac{0.00024 \times (A - B) \times 1000}{25}$ gm of Mg

10^6 ml of water contains = $\frac{0.00024 \times (A - B) \times 10^6}{25}$ gm of Mg.

Milliequivalents of magnesium/litre

$= \frac{0.00024 \times (A-B) \times 1000 \times 1000}{25 \times 24}$

$= \frac{(A - B) \times 10}{25}$

9. SODIUM AND POTASSIUM

Principle

In good quality waters, the amount of sodium and potassium present is very little. But in saline and brackish waters, the concentration of sodium may be quite high and the electrical conductivity of the waters will also be more than 1 mmhos/cm. The amount of sodium and potassium present is estimated with a flame photometer. Sodium emits a bright yellow colour (589 mμ) when excited in the flame. Similarly potassium also emits a radiation which can be read. The intensity of emission is proportional to the concentration of sodium and potassium in water.

Special equipment

Flame photometer

Procedure

Prepare a standard curve using sodium chloride (AR) of different concentrations. Then introduce the water sample in the flame photometer and find out the reading. From the standard curve, find out the concentration of sodium in the water sample.

Calculations

Milliequivalents of sodium /litre = ppm / equivalent weight
= ppm / 22.99
= ppm x 0.04348

Potassium

Principle

As indicated earlier, potassium gives a violet colour radiation at 404.4 and 767 mμ. The amount present in the water sample is usually estimated at 404.4 mμ with blue filter.

Special equipment

Flame photometer

Procedure

As in the case of sodium, prepare a standard curve for potassium using potassium chloride (AR) of different concentrations. Then introduce the water sample in the flame photometer and find out the reading. From the standard curve, find out the concentration of potassium in the water sample.

Calculations

Milliequivalents of sodium /litre = ppm / equivalent weight
= ppm / 39.10
= ppm x 0.0256

10. RESIDUAL SODIUM CARBONATE (RSC)

The residual sodium carbonate gives a measure of the sodium hazard. It is calculated using the amount of carbonate, bicarbonate, calcium and magnesium present in the waters.

$$\text{RSC (meq/litre)} = (CO_3^{2+} + HCO_3^-) - (Ca^{2+} + Mg^{2+})$$

RATING : <1.25	= Can be safely used
1.25 - 2.50	= Marginal, should be used with caution
> 2.50	= Unsuitable

11. WATER CLASS RATING

The irrigation waters are classified into various categories based on electrical conductivity and the calcium plus magnesium contents. The experimentally determined Ca + Mg values in meq/litre is spotted on the chart and a perpendicular is drawn from the point. Then the EC (electrical conductivity) of the water is noted on the chart and a slanting line is drawn exactly parallel to the line for next higher value. The point where the EC line intersects the perpendicular drawn from the calcium plus magnesium point would fall in a particular zone in the chart and that gives the water class. Now a perpendicular from the point of intersection is extended to sodium absorption ratio line

$$SAR = \frac{Na^{+}}{\sqrt{\frac{Ca^{2+} + Mg^{2+}}{2}}}$$

meeting it at some point. This point denotes approximately the sodium absorption ratio (SAR).

12. PERMANENT HARDNESS OF WATER

Principle

Salts of calcium which cause permanent hardness react with sodium carbonate forming insoluble calcium carbonate

$$Ca^{2+} + Na_2CO_3 \text{ -----> } CaCO_3 + 2\,Na^{+}$$

Similarly insoluble magnesium hydroxide is formed when sodium hydroxide reacts with magnesium salts.

$$Mg^{2+} + 2\,NaOH \text{ -----> } Mg(OH)_2 + 2\,Na^{+}$$

When an excess of standard solution of soda reagent is added, it will be consumed by the Ca^{2+} and Mg^{2+} ions. The amount of alkali consumed will indicate the extent of permanent hardness.

Reagents

Standard soda reagent (0.1 N), sulphuric acid (0.1 N), methyl orange indicator.

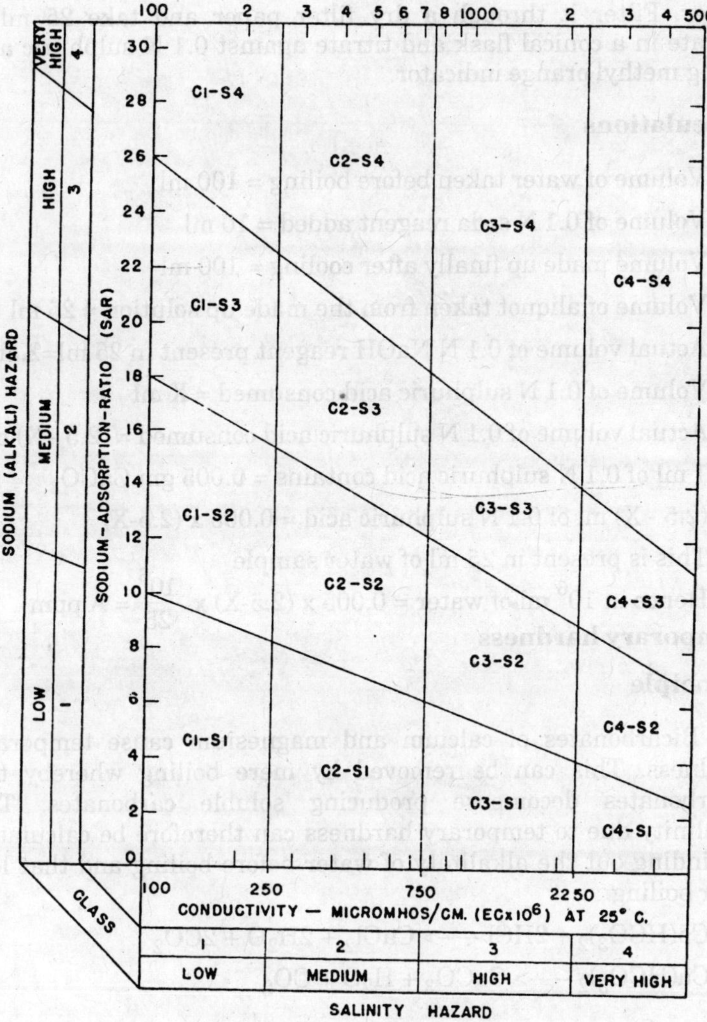

Classification of irrigation waters

Procedure

To 100 ml of water sample taken in a beaker, add 10 ml of soda reagent, boil for 15 minutes, cool and transfer to a 100 ml measuring flask and then make up to the mark with distilled water. Filter it through a dry filter paper and take 25 ml of filtrate in a conical flask and titrate against 0.1 N sulphuric acid using methyl orange indicator.

Calculations

Volume of water taken before boiling = 100 ml

Volume of 0.1 N soda reagent added = 10 ml

Volume made up finally after cooling = 100 ml

Volume of aliquot taken from the made up solution = 25 ml

Actual volume of 0.1 N NaOH reagent present in 25 ml=2.5ml

Volume of 0.1 N sulphuric acid consumed = X ml

Actual volume of 0.1 N sulphuric acid consumed = (2.5 - X) ml

1 ml of 0.1 N sulphuric acid contains = 0.005 gm $CaCO_3$

(2.5 - X) ml of 0.1 N sulphuric acid = 0.005 x (2.5-X)

This is present in 25 ml of water sample

Hence in 10^6 ml of water = $0.005 \times (2.5\text{-}X) \times \frac{10^6}{25}$ = A ppm

Temporary hardness

Principle

Bicarbonates of calcium and magnesium cause temporary hardness. This can be removed by mere boiling whereby the bicarbonates decompose producing soluble carbonates. The alkalinity due to temporary hardness can therefore be calculated by finding out the alkalinity of water before boiling and that left after boiling.

$$Ca(HCO_3)_2 + 2HCl \text{ -----> } CaCl_2 + 2H_2O + 2CO_2$$

$$Ca(HCO_3)_2 \text{ -----> } CaCO_3 + H_2O + CO_2$$

Reagents

Sulphuric acid (0.1 N), methyl orange

Procedure

Take 50 ml of the water sample in a 250 ml conical flask and titrate against 0.1 N sulphuric acid using two drops of methyl orange as indicator. The end point is the appearance of pink colour. This will give the total carbonate hardness.

Then take another 100 ml of water sample in a beaker and evaporate to dryness, cool it and treat it with 50 ml of distilled water. Warm and stir to dissolve the water soluble portion of the residue, filter it into a 100 ml measuring flask and make up to the volume. Take 50 ml of this solution and titrate it to get the non-bicarbonate hardness.

Calculations

A. Volume of water sample taken before boiling = 50 ml

Volume of 0.1 N sulphuric acid required for neutralisation = V_1 ml

1 ml of 0.1 sulphuric acid = 0.005 gm $CaCO_3$

V_1 ml of 0.1 sulphuric acid = 0.005 x V_1 gm of $CaCO_3$

This is present in 50 ml of water

Hence in 10^6 ml of water = $0.005 \times V_1 \times \frac{10^6}{50}$ ppm

This gives the total carbonate hardness in ppm = X ppm

B. Volume of water sample taken for boiling = 100 ml

Volume of water extract taken for titration = 50 ml

Volume of 0.1 N sulphuric acid used for neutralisation = V_2 ml

0.1 N sulphuric acid = 0.005 gm $CaCO_3$

V_2 ml of 0.1 sulphuric acid = 0.005 x V_2 gm of $CaCO_3$

In 10^6 ml of water = $0.005 \times V_2 \times \frac{10^6}{50} \times 100$

= Y ppm

Temporary hardness = (X- Y) ppm $CaCO_3$

CLASSIFICATION OF WATERS ACCORDING TO DEGREE OF HARDNESS

Classification	Hardness (ppm)	Classification	Hardness (ppm)
Extremely soft	< 15	Hard	110-130
Very soft	15 - 30	Very hard	130-170
Moderately soft	30 - 90	Excessively hard	170-230
Moderately hard	90 - 110	Too hard	>230

STANDARDS OF QUALITY OF WATER FOR VARIOUS INDUSTRIES

(Limiting tolerances of impurities in ppm)*

Impurity	Washing	Irrigation	Textiles	Sugar factory	Pottery	Steam Boiler
Ca^{2+}	10	40	10	20	–	5
Mg^{2+}	5	20	5	10	–	5
Cl^-	500	100	100	20	20	100
SO_4^{2-}	500	200	100	20	20	100
Fe^{2+}	0.1	–	0.1	0.1	–	–
HCO_3	–	–	200	100	–	100

*** As these are changed as and when fresh information is generated, please check up for the latest standards.**

CHAPTER - 10

ANALYSIS OF FERTILIZERS AND MANURES

Most of the fertilizers are chemically pure compounds containing one or more of the macronutrients (nitrogen, phosphorus or potassium) and in certain cases one or two other nutrients like sulphur, copper, zinc and magnesium. Earlier the fertilizers and manures were analysed for the total nutrient content but now in most of the cases, the individual forms in which the nutrient is present, are also estimated. For example in nitrogenous fertilizers, the total nitrogen content as well as ammonical nitrogen or nitrate-nitrogen content are also estimated. Similarly in phosphatic fertilizers water soluble phosphorus, citrate soluble phosphorus, and citrate insoluble phosphorus are also estimated in addition to the total phosphorus. Manures and manure mixtures contain considerable amount of organic matter and other materials and hence they have to be analysed only after converting the nutrients into the inorganic form by some pretreatment. Different methods of analysis of the fertilizers and manures commonly used are given below in detail for the estimation of many of the common macronutrients, secondary nutrients and micronutrients. Certain of the procedures are now currently in use in certain of the fertilizer quality control laboratories also.

A. NITROGENOUS FERTILIZERS

1. Total nitrogen in ammonical fertilizers (Ammonium

sulphate, ammonium chloride) and ammonium-nitrogen in ammonium nitrate and calcium ammonium nitrate.

Principle

The ammonical nitrogen present in the fertilizer is distilled directly by reacting it with alkali and then the liberated ammonia is absorbed in standard acid and then estimated.

$$(NH_4)_2 SO_4 + 2NaOH \text{ ---- } Na_2SO_4 + 2\,NH_3 + 2\,H_2O$$

Reagents

Sulphuric acid (0.1 N), potassium hydroxide (0.1 N), sodium hydroxide (40%), freshly prepared (ignited) carbonate-free magnesium oxide, methyl red indicator.

Procedure

Weigh accurately about 1 gm of the fertilizer and transfer it to a 250 ml beaker. Add 50 ml of distilled water and dissolve the same and transfer it to a 250 ml volumetric flask, washing the beaker with small amounts of water. Make up to the volume with distilled water (In case there is any insoluble material, filter the solution in the beaker through Whatman No. 1 filter paper into a 250 ml volumetric flask).

Distillation

Pipette 10 ml of the extract into a distillation flask, add 400 ml of distilled water, few glass beads, 2 or 3 zinc bits and close the flask.

Simultaneously pipette out exactly 25 ml of 0.1 N sulphuric acid into an ice tumbler or 250 ml tall form beaker, add 2 or 3 drops of methyl red indicator and a small amount of distilled water (20-40 ml). Keep the ice tumbler with the contents below the delivery tube dipping inside the acid solution but not touching the bottom of the container.

After setting up the delivery end of the flask, open the distillation flask, add 10 ml of 40% sodium hydroxide to the flask and close it tightly (Instead of the alkali, 2 gm of freshly ignited carbonate free magnesium oxide can also be used).

Heat the distillation flask slowly initially and then steadily. Continue till all the ammonia has distilled over. This can be tested by keeping a red litmus near the delivery end of the distillation flask. If it does not turn blue, then it indicates that no more ammonia is distilling over. However if distillation is continued beyond this point, alkali may start distilling and this will give blue colour to the litmus. Hence this test should be carried out early.

As soon as the distillation is over, remove the ice tumbler containing 0.1 N sulphuric acid, wash the delivery tube with distilled water, collecting the washings in the same ice tumbler and back titrate the excess acid with 0.1 N potassium hydroxide, till the appearance of the straw yellow colour. The difference gives the volume of 0.1 N acid actually neutralised by the ammonia. From this, the amount of nitrogen present in the fertilizer can be calculated.

Calculations

Weight of fertilizer sample taken = W gm

Volume of extract made up = 250 ml

Volume of aliquot taken for analysis = V ml

Volume of 0.1 N sulphuric acid taken in the ice tumbler = 25 ml

Volume of 0.1 N potassium hydroxide used in back titration = X ml

Volume of 0.1 N sulphuric acid actually consumed to neutralise the evolved ammonia = (25 - X) ml

1 ml of 0.1 sulphuric acid = 0.0014 gm N

(25 - X) ml of 0.1 N sulphuric acid = 0.0014 x (25-X) gm N

This is found in "V" ml of the extract.

Hence in 250 ml of extract (equivalent to "W" gm of fertilizer) $= 0.0014 \times (25 - X) \times \frac{250}{V}$

Per cent of ammonical nitrogen

$$= 0.0014 \times (25 - X) \times \frac{250}{V} \times \frac{100}{W} \text{ gram of nitrogen}$$

Precautions

1. During the distillation, in case the acid in the ice tumbler turns yellow, it indicates that all the acid has been neutralised by the ammonia. In that case, add another aliquot (exact amount) of the 0.1 N sulphuric acid to the ice tumbler without disturbing the set up and then continue the distillation. Record the additional amount of standard acid added and modify the calculations accordingly.

2. Any sudden change particularly a decrease in gas pressure and consequently the heat will result in the back suction of the solution from the ice tumbler. Hence the steady heating of the distillation flask has to be maintained throughout the period of distillation.
3. In case of failure of gas supply or a decrease in heat due to other causes, disconnect the delivery tube connecting the distillation set to the ice tumbler immediately on noticing signs of back suction. Restore the gas supply, reconnect the tube and continue distillation. If possible repeat the estimation.
4. This method assumes that urea is absent in the sample.
5. Also the method cannot be applied in case of compounds like $MgNH_4PO_4$ or $FeNH_4 PO_4$.

2. NITRATE - NITROGEN (NO_3 - N)

(Devarda's alloy method) (for sodium nitrate, calcium nitrate, ammonium nitrate, potassium nitrate)

Principle

This method estimates the nitrate-nitrogen present in the fertilizers. The Devarda's alloy that is added liberates the nascent hydrogen and this reduces the nitrate-nitrogen to ammonia which is estimated by absorbing the same in acid and titrating the excess as before.

$$Zn + 2\ NaOH \rightarrow Na_2\ ZnO_2 + 2\ (H^+)\ \text{Sodium zincate}$$

$$NO_3^- + 9\ (H^+) \rightarrow NH_3 + 3\ H_2\ O$$

Reagents

Sulphuric acid (0.1 N), potassium hydroxide (0.1 N), sodium hydroxide (40%), Devarda's alloy composite, methly red indicator.

Procedure

Weigh accurately a known amount of fertilizer (about 1 gm) and transfer it to a clean beaker. Add small amount of distilled water, dissolve the fertilizer and transfer it to a 250 ml volumetric flask carefully (If there are any insoluble materials, transfer it through a Whatman No. 1 filter paper). Wash the beaker number of times with distilled water and transfer it to the volumetric flask carefully and then make upto the mark with distilled water.

Then pipette out 25 ml of 0.1 N sulphuric acid into the ice tumbler (or beaker), add 3 to 5 drops of methyl red indicator and place it at the delivery end of the distillation flask.

Distillation

Pipette out 25 ml of the fertilizer extract into the distillation flask, add 400 ml of water, 3 gm of Devarda's alloy, few glass beads and 5 ml of 40% sodium hydroxide. Close the flask immediately and allow the reaction to proceed in the cold for a short period and then gently heat the flask.

Collect the distillate in the ice tumbler till the distillate is free of ammonia as observed by the red litmus test, back titrate the excess of 0.1 N sulphuric acid against 0.1 N potassium hydroxide till the appearance of straw yellow colour. From the volume of 0.1 N acid consumed, calculate the amount of nitrate-nitrogen present in the sample.

Calculations

Weight of fertilizer taken = W gm

Volume of fertilizer extract made up = 250 ml

Aliquot of extract taken = V ml

Volume of 0.1 N sulphuric acid taken in the ice tumbler = 25 ml

Volume of 0.1 N potassium hydroxide used in back titration= X ml

Volume of 0.1 N sulphuric acid consumed = (25 - X) ml

1 ml of 0.1 N sulphuric acid = 0.0014 gm of N

(25 - X) ml of 0.1 N sulphuric acid = 0.0014 x (25-X) gm of N

This is present in "V" ml of extract

Amount of NO_3-N present in 250 ml $= 0.0014 \times (25 - X) \times \frac{250}{V}$

(This is equivalent to W gm of fertilizer)

Per cent of (NO_3 - N) in the fertilizer

$$= 0.0014 \times (25 - X) \times \frac{250}{V} \times \frac{100}{W}$$

Precautions

Observe the same precautions (1 to 3) given in the ammonical - nitrogen estimation.

3. FERTILIZERS CONTAINING BOTH FORMS OF NITROGEN

a. When the nitrogen fertilizer contains both ammonical and nitrate-nitrogen as in ammonium nitrate and when the amount of each form in the fertilizer has to be estimated, then prepare the fertilizer extract as indicated in the estimation of either ammonical nitrogen or nitrate nitrogen.

b. Then pipette out a known volume of the extract into the distillation flask and distill with 10 gm of freshly ignited magnesium oxide as in the estimation of ammonical nitrogen and estimate the amount of nitrogen present.

c. After completing the estimation, cool the contents of the distillation flask and add 4 gm of Devarda's alloy and 10 ml of 40% sodium hydroxide as in the nitrate-nitrogen estimation and estimate the amount of nitrogen present in the sample.

d. Calculations are same as given in the estimations (1) and (2).

4. NITROGEN IN THE AMIDE FORM (Urea, Calcium cyanide)

Principle

Urea contains nitrogen in the amide (organic) form. Hence it has to be converted into inorganic form by reacting it with strong acid.

$$CO(NH_2)_2 + 2H_2SO_4 \text{ -----> } (NH_4)_2SO_4 + SO_2 + CO_2 + H_2O$$

The ammonium sulphate that is formed is distilled with alkali as in the case of ammonical nitrogen and the amount of nitrogen determined as before.

Reagents

Concentrated sulphuric acid, copper sulphate, potassium sulphate, sodium hydroxide (40%), sodium sulphide (10%), sulphuric acid (0.1 N), potassium hydroxide (0.1 N), methyl red indicator.

Procedure

Weigh accurately a known quantity of the urea sample (about 1 gm), transfer it to a 250 ml volumetric flask, add water, dissolve it and make upto the mark.

Digestion

Pipette out 10 ml of this extract into a Kjeldahl flask, add 20 ml of concentrated sulphuric acid, 5 gm of copper sulphate and 5 gm of potassium sulphate. Digest it in a fume cupboard till the contents are clear and greenish blue in colour. Cool the Kjeldahl flask.

Distillation

Transfer the contents of Kjeldahl flask to the distillation flask. Then rinse the Kjeldahl flask with water and transfer the same to the distillation flask and repeat it number of times (4 to 5 times).

Then add a few glass beads, 100 ml of 40% sodium hydroxide and 10 ml of sodium sulphide (Na_2S). Close the flask and distill the contents and absorb the evolved ammonia in the 0.1 N sulphuric acid (plus methyl red) kept in the ice tumbler.

When the distillate is free of ammonia, disconnect the delivery tube, wash it with water into the ice tumbler and back titrate the excess acid with the 0.1 N potassium hydroxide, till the appearance of the straw yellow colour. From this calculate the amount of nitrogen present in the fertilizer.

Calculations

Weight of fertilizer = W gm

Volume of extract made up = 250 ml

Volume of extract taken for analysis = V ml

Volume of 0.1 N sulphuric acid taken in the ice tumbler = 25 ml

Volume of 0.1 N potassium hydroxide used in back titration=X ml

Volume of 0.1 N sulphuric acid neutralised by the ammonia
= (25-X) ml

1 ml of 0.1 N sulphuric acid = 0.0014 gm N

(25 - X) ml of 0.1 N sulphuric acid = 0.0014 x (25-X)

This is found in "V" ml of the extract

Amount of nitrogen in 250 ml of extract (= W gm fertilizer)

$$= 0.0014 \times (25 - X) \times \frac{250}{V}$$

Per cent of nitrogen in the urea sample

$$= 0.0014 \times (25 - X) \times \frac{250}{V} \times \frac{100}{W}$$

5. TOTAL NITROGEN IN MANURES

(Farm yard manure, compost, groundnut cake)

Principle

The organic form of nitrogen in the manures is converted into ammonium sulphate by digestion with concentrated sulphuric acid by the salicylic acid method. The nitrates present react with the sulphuric acid to yield nitric acid.

$$2NaNO_3 + H_2SO_4 \text{ -----> } 2\ HNO_3 + Na_2SO_4$$

The nitric acid formed in the presence of concentrated sulphuric acid reacts with the added salicylic acid and forms nitrosalicylic acid

$$C_6H_4\ OHCOOH + HNO_3 \text{ -----> } \underset{\text{Nitrosalicylic acid}}{C_6\ H_3\ OHCOOHNO_2} + H_2O$$

The nitrosalicylic acid is reduced to aminosalicylic acid by sulphuric acid produced by the action of sulphuric acid and sodium thiosulphate.

$$Na_2\ S_2\ O_3 + H_2\ SO_4 \text{ ------> } Na_2\ SO_4 + \underset{\text{Hydrogen sulphite}}{H_2\ SO_3} + S$$

$$C_6H_30HCOOHNO_2{+}3\ H_2SO_3{+}H_20 \text{ ->} \underset{\text{Aminosalicylic acid}}{C_6H_3NH_20HCOOH}{+}3H_2SO_4$$

Then the aminosalicylic acid is oxidised by sulphuric acid and the nitrogen is converted to ammonium sulphate.

$$2\ C_6H_3NH_20HCOOH + 27\ H_2SO_4 \text{ ------> } (NH_4)_2\ SO_4 + 26\ SO_2 + 14\ CO_2 + 30\ H_20$$

The ammonium sulphate that is formed is estimated by direct distillation with excess alkali and the ammonia evolved is absorbed in the acid and estimated.

Reagents

Concentrated sulphuric acid, salicylic acid, sodium thiosulphate, potassium sulphate, copper sulphate, sulphuric acid (0.1 N), potassium hydroxide (0.1 N), methyl red indicator, sodium hydroxide (40%), sodium sulphide (10%).

Procedure

Transfer a known weight of the representative sample of the manure (about 1 gm) and add 30 ml of concentrated sulphuric acid and 1 gm of salicylic acid. Mix the contents well and allow it to stand at room temperature for 30 minutes with frequent shaking. Then add about 5 gm of sodium thiosulphate and shake it well. Digest it over a low flame till frothing ceases. Then add 10 gm of potassium sulphate and 1 gm of copper sulphate and digest it over a strong flame until the liquid turns green. Then cool the flask to the room temperature.

Distillation

Transfer the contents of the Kjeldahl flask to the distillation flask. Wash the Kjeldahl flask by repeated washings into the distillation flask. Add few porcelain and zinc bits and 100 ml of 40% sodium hydroxide and 10 ml of 10% sodium sulphide.

Before starting the distillation, keep an ice tumbler (or tall form beaker) with 25 ml of 0.1 N sulphuric acid and few drops of methyl red indicator at the delivery end of the distillation flask. Then start the distillation and absorb the evolved ammonia in the acid and after completion of the distillation (evolution of ammonia) as tested by the litmus paper test, disconnect the delivery tube and wash it with water into the ice tumbler. Back titrate the excess acid with 0.1 N potassium hydroxide till the colour of the solution turns yellow. From this calculate the amount of nitrogen present in the sample.

Calculations

Weight of manure = W gm

Volume of 0.1 N sulphuric acid = 25 ml

Volume of 0.1 N potassium hydroxide used = X ml

Volume of 0.1 N sulphuric acid neutralised by the evolved ammonia = (25 - X) ml

1 ml of 0.1 N sulphuric acid = 0.0014 gm N

(25 - X) ml of 0.1 N sulphuric acid = 0.0014 x (25-X) gm N

This is present in "W"" gm of the sample

$$\text{Per cent of nitrogen in the sample} = 0.0014 \times (25\text{-}X) \times \frac{100}{W}$$

6. NEUTRAL UREASE METHOD
(for nitrogen in urea in mixed fertilizers)

Principle

The nitrogen is estimated by simple volumetric method using

neutral urease solution. This method is recommended to be followed in the fertilizer quality control laboratories by the Fertilizer control order.

Reagents

Neutral urease solution, saturated barium hydroxide, sodium carbonate (10%), hydrochloric acid (2 N), sodium hydroxide (0.1 N), hydrochloric acid (0.1 N), methyl purple indicator.

Procedure

Transfer a known quantity of fertilizer (10 gm) and dissolve it in a small amount of water and transfer it to a 500 ml volumetric flask through Whatman No. 1 filter paper. Leach it number of times and then add 75 - 100 ml of saturated barium hydroxide solution to precipitate the phosphates. Allow it to settle and test for the completion of precipitation with a few drops of barium hydroxide solution. Add 20 ml of 10% sodium carbonate solution to precipitate the excess barium and any soluble calcium salts. Allow it to settle and test for the completion of precipitation. Dilute to volume, mix and filter again through Whatman No. 12 filter paper. Transfer 50 ml of this solution to 250 ml conical flask and add 1 to 2 drops of methyl purple indicator. Acidify the solution with 2 N hydrochloric acid and add 2 to 3 drops in excess. Neutralise the solution with 0.1 N sodium hydroxide to just first change of the colour of the indicator. Add 20 ml of neutral urease solution, close the flask with rubber stopper and let it stand for 1 hour at 20 - 25°C. Cool the flask in ice slurry and titrate at once with 0.1 N hydrochloric acid to full purple colour, then add approximately 5 ml excess. Note the total volume added, back titrate excess hydrochloric acid with 0.1 N sodium hydroxide to neutral end point.

Calculations

a. Per cent of urea =

$$\frac{(\text{ml of 0.1 N HCl - ml of 0.1 N NaOH}) \times 0.3003 \text{ times}}{\text{weight of the sample}}$$

b. Total nitrogen in samples containing nitrates : Digest the sample by the salicylic acid method and estimate the same by distilling the ammonia and collecting it in the acid.

c. Total nitrogen content in materials with high Cl : NO_3 ratio and those with high water soluble nitrogen: Digest accurately weighed quantity of the fertilizer with 2 to 5 gm of reduced Fe, 25 ml of cold 1:1 sulphuric acid, first at room temperature till the reaction ceases. Then add boiling chips and boil for 15 to 20 minutes and then cool, and distill as in the other method. In case the sample contains organic matter, after boiling with sulphuric acid (1:1), add 0.7 gm of mercuric acid (HgO) and boil again for 40 minutes, cool and then distill.

d. In case of estimation of nitrate - nitrogen in presence of amide forms of nitrogen in the mixture (urea or calcium cyanamide), take a known quantity of the fertilizer extract in 500 ml Kjeldahl flask, add 2 gm of $FeSO_4.7H_2O$ and 20 ml of concentrated sulphuric acid and digest it for 10 more minutes after the white fumes appear, then add 0.65 gm of mercury or 0.79 gm of mercuric oxide (HgO) till all organic matter is oxidised. Cool, dilute and add potassium sulphide solution and determine the nitrogen as described earlier. To prevent bumping in the distillation flask, add a pinch of zinc dust and zinc granules to the flask.

e. The water insoluble nitrogen in any sample can be estimated by repeatedly leaching the samples with water through Whatman filter paper No. 42 and then collecting the residue on the filter paper and estimating the nitrogen by digestion and then by distillation.

7. FREE ACIDITY IN NITROGENOUS FERTILIZERS

Principle

The fertilizer is dissolved in a known amount of distilled water and titrated against a standard alkali. From this the amount of free acidity is calculated.

Reagents

Sodium hydroxide (0.02 N), methyl red indicator, or methyl red - methylene blue mixed indicator.

Procedure

Dissolve 20 gm of the fertilizer sample in 50 ml of cold water, filter and make upto 200 ml volume in a volumetric flask. Titrate this solution against 0.02 N sodium hydroxide using methyl red or methyl red - methylene blue mixed indicator.

Calculations

Weight of fertilizer taken = W gm

Volume of standard sodium hydroxide used = V ml

Normality of sodium hydroxide = 0.02 N

Free acidity as sulphuric acid per cent by weight

$$= \frac{4.904 \times V \times 0.02}{W}$$

8. BIURET CONTENT IN NITROGENOUS FERTILIZERS

Principle

Presence of biuret in nitrogenous fertilizers has been found to have a phytotoxic effect if it exceeds a certain limit. Hence the biuret present in the sample is extracted with an ion exchange resin and estimated colorimetrically at 555 mμ.

Reagents

Alkaline tartarate solution, copper sulphate (1.5%), biuret standard solution (1 mg/ml), sulphuric acid (0.1 N), methyl red indicator.

Special equipments

Ion exchange resin, colorimeter with a filter for 500-570 mμ region.

Procedure (In urea samples)

Dissolve 2 to 5 gm of urea sample in 100 ml water and keep it in a water bath at 50°C for 30 minutes, filter through Whatman No. 1 filter paper into a 250 ml volumetric flask, wash it number of times and make up to the mark.

i. Take 25 ml of the solution in a 100 ml volumetric flask and add a drop of methyl red indicator and carbon dioxide free water to make up to 50 ml. Neutralise it with 0.1 N sulphuric acid till the solution becomes pink and add 20 ml of alkaline tartarate and 20 ml of copper sulphate solutions. Dilute to the volume, shake it for 10 seconds and place it on a water bath for 15 minutes at 30 ± 5°C.

Simultaneously prepare a blank and determine the absorbance of the solutions at 555 mμ with 500- 570 mμ filter and note the reading. From the standard curve, find out the amount of biuret present.

In case of mixed fertilizers, stir an accurately weighed fertilizer sample (10-20 gm) in 150 ml of hot water (50°C) for three minutes, filter and wash into a 250 ml volumetric flask and make up to the volume.

Transfer 25 ml aliquot to an ion exchange resin column and when the liquid level drops to the top of the resin bed, wash with two 25 ml aliquots of water and collect the elute in the same beaker. To the combined elutes, add 2 drops of methyl red and then 0.1 N sodium hydroxide to bring it to the yellow colour. Add 0.1 N sulphuric acid till the solution becomes pink and transfer this to 100 ml volumetric flask and dilute with carbon dioxide free water.

Transfer 50 ml of the above solution to a 100 ml volumetric flask and proceed as in (i) of this section.

For finding out the biuret content in the fertilizer, determine the concentration of biuret in the final dilution using the standard curve. Then calculate the biuret content.

$$\text{Per cent biuret} = \frac{C \times 100}{W}$$

Where C = concentration of biuret in mg/ml in the final dilution from the standard curve

W = concentration of original sample in final dilution expressed as mg/ml.

B. PHOSPHATIC FERTILIZERS

1. TOTAL PHOSPHORUS (in superphosphate, triple superphosphate)

Principle

The total amount of phosphorus in the fertilizer is extracted with dilute acid and then it is precipitated as ammonium phosphomolybdate which is determined volumetrically. From this the amount of P_2O_5 present in the fertilizer is calculated.

Reagents

Nitric acid (0.1619 N), potassium hydroxide (0.1619 N), solid ammonium chloride, solid ammonium nitrate, phenolphthalein indicator, ammonium molybdate (20%), precipitating mixture (10 ml of 20% ammonium molybdate plus 7 ml of concentrated nitric acid and 3 ml of water), dilute hydrochloric acid.

Procedure

a. Transfer an accurately weighed quantity (1 to 2 gm) of fertilizer to a 150 ml beaker, add about 30 ml of dilute hydrochloric acid (1:1) and cover the beaker and digest it for

30 minutes over a sand bath. Cool the contents and filter it through Whatman No. 40 filter paper and wash it with hot water till the washings are free of chloride (test with silver nitrate). Collect the filtrate in 250 ml volumetric flask, cool and then make up to the mark.

b. Estimation of phosphorus: Pipette out 10 ml of the extract into a 250 ml beaker, add a piece of red litmus, then add dilute ammonia drop by drop till the contents are alkaline. Then add concentrated nitric acid till the solution is acidic as seen by the litmus paper. To this add 10 gm of solid ammonium nitrate and place it in a thermostat at 65°C for 10 minutes. To this, add slowly 20 ml of freshly prepared precipitating mixture, stir well with a glass rod without touching the sides of the beaker till the canary yellow precipitate completely settles down leaving a clear supernatent liquid.

Filter the solution through Whatman No. 40 filter paper transferring only the supernatent liquid to the filter paper. Wash the precipitate 6 - 8 times in the same manner till the filtrate is free of acid. Do not allow either the beaker or the filter paper to become dry during the washings. After 6 to 8 washings, test the filtrate to see whether it is free of acid.

c. Test for free of acidity : Add a drop of 0.1619 N potassium hydroxide and phenolphthalein to a test tube and collect at least 15 ml of the filtrate into the test tube. Shake well and if the pink colour is not discharged, it indicates that the filtrate is free of acid.

After the filtrate is free of acid, remove the filter paper with the precipitate to the original beaker in which the precipitation was carried out. Add 100 to 150 ml of distilled water through the sides of the beaker and pulp the filter paper using a glass rod. Then add a known amount of 0.1619 N potassium hydroxide from the burette till the precipitate completely dissolves as seen by the disappearance of yellow colour. Add about 5 to 10 ml of alkali in excess and then a drop or two of phenolphthalein indicator. The solution will be pink in colour and titrate this against 0.1619 N nitric acid until the pink colour disappears. From the titre value of 0.1619 N potassium hydroxide consumed, calculate the amount of P_2O_5 present in the sample.

Calculations

Weight of fertilizer taken = W gm

Volume of extract made up = 250 ml

Volume of extract taken for estimation = V ml

Volume of 0.1619 N KOH added = X ml

Volume of 0.1619 N HNO_3 used for back titration = Y ml

Volume of 0.1619 N KOH = (X-Y) ml

1 ml of 0.1619 N KOH = 0.0005 gm of P_2O_5

(X-Y) ml of 0.1619 N KOH = 0.0005 x (X-Y) gm of P_2O_5

This is present in "V" ml.

Amount of P_2O_5 present in 250 ml = $0.0005 \times (X\text{-}Y) \times \frac{250}{V}$ gm of P_2O_5.

This is present in "W" grams.

Hence per cent of total P_2O_5 in the fertilizer

$$= 0.0005 \times (X\text{-}Y) \times \frac{250}{V} \times \frac{100}{W}$$

Per cent of total P in the fertilizer = Total P_2O_5 x 0.43

Precautions

1. Do not allow the precipitate or the filter paper to dry even for a short time. If it dries, molybdic acid may start forming. Once it starts forming, it will be extremely difficult to free the precipitate of the acid. Even if one succeeds, the value may not reflect the correct end point. In such cases it is better to repeat the analysis.
2. The yellow precipitate has always got a tendency to creep up in the filter paper and pass through the pores between the filter paper and the glass funnel. Hence while washing the precipitate, the solution should not be transferred to more than 3/4th of the filter paper.

ii. Total phosphorus in farm yard manure/compost/ groundnut cake (by volumetric method)

In the case of organic manures, digest 1 gm of sample with 12 ml of triple acid (9:2:1 of nitric acid : sulphuric acid : perchloric acid) over sand bath, until a clear white residual solution is obtained. After digestion, transfer it to a 100 ml of volumetric flask and make up to the mark.

From the above solution, pipette out 25 ml of the triple acid extract and estimate P_2O_5 as in the case of fertilizers (superphosphate - section "b")

iii. Colorimetric method of estimation of phosphorus in triple acid extract

Principle

The triple acid extract is treated with Barten's reagent and the colour developed is measured in a photoelectric colorimeter using blue filter (420 mμ)

Reagents

Barten's reagent

Special equipment

Colorimeter

Procedure

Pipette out 5 ml of the triple acid extract into 25 ml volumetric flask and add 5 ml of Barten's reagent and make upto the volume. Allow it to stand for 30 minutes for the development of yellow colour and measure the intensity of colour using a photoelectric colorimeter with a blue filter (420 mμ) after adjusting the transmittance to 100 with a blank. The colour is stable for 24 hours. From the standard curve, find out the amount of P in the solution and calculate the percentage of P_2O_5 in the manure.

Calculations

Weight of manure sample taken = W gm

Volume of triple acid extract prepared = 100 ml

Volume of triple acid extract taken for analysis = V ml

Volume of solution made up = 25 ml

Concentration of P in the solution as observed from the standard curve = X ppm

Concentration of P in the whole solution = $X \times \frac{100}{V}$ ppm

This is present in "W" gm of the sample

So in 100 gm of the sample = $X \times \frac{100}{V} \times \frac{100}{W}$ ppm

Per cent of P in the manure = $X \times \frac{100}{V} \times \frac{100}{W} \times \frac{25}{1000000}$

Convert this into P_2O_5 by multiplying the value by 2.29 if the standard curve has been drawn with concentrations expressed as P.

Precautions

In the preparation of the standard curve, the concentration of the solution is expressed in ppm of P or P_2O_5. A common mistake committed is to prepare the standard curve in terms of P and overlooking this fact, expressing the final result as so many grams of "P_2O_5" or vice versa without using the conversion factor. This should be avoided.

10. WATER SOLUBLE PHOSPHORUS IN FERTILIZERS

Principle

The fertilizer is leached with water and the amount of P in the leachate is estimated by the same procedure as in the estimation of total phosphorus.

Reagents

Same as in total P_2O_5 estimation.

Procedure

Weigh accurately a known amount of superphosphate and transfer it to a Whatman No. 40 filter paper fitted to a funnel. Wash the sample with successive small portions of distilled water allowing each portion to completely drain before adding the next portion. Collect the washings in a 250 ml measuring flask. Continue the leachings till the leachate is free of acid. Then make up to the volume and estimate the P_2O_5 as in the case of total phosphorus. If the filtrate is turbid, add a few drops of concentrated nitric acid and then take 10 ml of the leachate for estimation. Estimation and calculations are similar to that in total P_2O_5.

11. TOTAL PHOSPHORUS IN FERTILIZER
(Gravimetric Quinoline - molybdate method)

Principle

The phosphorus is extracted with diacid and the extracted phosphorus is precipitated with the quimaciac reagent as $(C_9H_7N)_3\ H_3PO_4\ 12MoO_3$ and weighed. This is multiplied by 0.03207 to get the weight of P_2O_5 in the aliquot.

Reagent

Citric molybdic acid reagent, quinoline solution, quimaciac reagent.

Procedure

Accurately weigh about one gram of the fertilizer, transfer it to a suitable flask and add 20 -30 ml of concentrated nitric acid, and 10 - 20 ml of 70 - 72 % perchloric acid. Boil gently for 30 - 45 minutes until the solution is colourless or nearly so and white fumes appear in the flask. Do not allow it to dry. Cool slightly, add 50 ml of quimaciac reagent and boil for a few minutes. Then transfer it to 250 ml volumetric flask and make up to volume.

Pipette out the aliquot of the solution into a 500 ml conical flask, dilute it with water to 100 ml and add 30 ml of citric molybdate acid reagent, boil gently for 3 minutes (solution must be precipitate-free at this stage) and then swirl carefully. Immediately add from the burette 10 ml of quinoline solution, initially adding drop wise up to 3 - 4 ml and then in a steady stream. Cool this to room temperature, mix thoroughly, filter in to gooch crucible with glass fibre filter paper previously dried at 250°C, weigh, wash 4 - 5 times with 25 ml portions of water. Dry the crucible and contents for 30 minutes at 250°C and cool it in a desiccator to constant weight, weigh the precipitate as $(C_9H_7N)_3$ H_3 PO_4. $12MoO_3$, subtract the weight of the reagent blank. Multiply the weight of the precipitate by 0.03207 to obtain the weight as P_2O_5. From this calculate the percentage of P_2O_5.

Or pipette out an aliquot of the diacid extract, dilute it to 100 ml with water and add 50 ml of quimaciac reagent, cover it with a watch glass, boil it on a hot plate for a minute, cool it to the room temperature. After thoroughly mixing, filter it through a gooch crucible with a glass fibre paper previously dried at 250°C and weighed. Wash the precipitate 5 times with 25 ml portions of water, dry the crucible and the contents for 30 minutes at 250°C. Cool in a desiccator to constant weight, subtract the weight of blank and find out the weight as $(C_9H_7N)_3$ H_3 PO_4. $12MoO_3$. Multiply the weight of the precipitate by 0.03207 to get the weight as P_2O_5 and from this calculate the percentage of P_2O_5.

(The digestion procedure can be simplified depending upon the nature of the constituent particularly organic matter) :

i. In case of materials containing small quantities of organic matter, dissolve the weighed quantity of fertilizer in 30 ml nitric acid and 3 - 5 ml hydrochloric acid, boil until the organic matter is destroyed.

ii. In case of materials containing high amount of organic matter, boil with 20 - 30 ml concentrated sulphuric acid and 2 - 4 gm of sodium nitrate or potassium nitrate. After the solution has become colourless, add another small quantity of the nitrate. When the solution is colourless, cool, add 150 ml of water and boil again. Before adding sodium nitrate or potassium nitrate, digest the mixture in a gentle heat until the violent reaction is over, then cool and add the nitrate.

iii. If the fertilizer contains much of Fe or Al phosphates or basic slag, dissolve the fertilizer in 15 - 30 ml concentrated hydrochloric acid and 3 - 10 ml of concentrated nitric acid.

iv. In case of organic materials like cotton seed meal or mixture, evaporate the materials with 5 ml of magnesium nitrate solution, ignite and then dissolve it in concentrated hydrochloric acid.

12. WATER SOLUBLE PHOSPHORUS BY GRAVIMETRIC METHOD

Procedure

Leach an accurately weighed quantity of fertilizer with small portions of water till it is free of acid as indicated in the volumetric estimation of water soluble phosphorus. Collect the leachates and make up to a known volume (250 ml) and pipette out an aliquot into a 500 ml erlenmeyer flask, dilute with water, add 10 ml of nitric acid (1 : 1) and boil gently for 10 minutes. Cool and dilute and estimate using the quimaciac reagent.

13. CITRATE INSOLUBLE PHOSPHORUS

After extracting all the water soluble phosphorus, transfer the residue with the filter paper to a 250 ml conical flask containing 100 ml of ammonium citrate solution previously heated to 65°C. Close the flask tightly and shake vigourously till the paper is reduced to a pulp and agitate the contents of the flask at exactly 65°C for one hour. Shaking should be such that the materials are dispersed in the citrate solution throughout. After shaking, filter the contents by suction through Whatman No. 5 filter paper using buchner funnel. Wash with distilled water and at 65°C, until the volume of the filtrate is approximately 350 ml, allowing time for thorough draining before adding more water. If the material is turbid, wash it with 2% ammonium nitrate solution. Determine the P_2O_5 content in the citrate insoluble residue either (i) by igniting all the organic matter till it is destroyed and digesting it with 10-15 ml concentrated hydrochloric acid till all the phosphate dissolves or (ii) by treating the wet precipitate and filter as in the case of total P_2O_5 digestion.

14. CITRATE SOLUBLE PHOSPHORUS

The citrate soluble phosphate is given by subtracting the sum of water soluble and citrate insoluble P_2O_5 from the total P_2O_5.

15. CITRIC ACID SOLUBLE PHOSPHORUS

Principle

A known quantity of the fertilizer is shaken with citric acid (2%) for 30 minutes and the amount of phosphorus extracted is estimated by the methods described earlier. This is supposed to give the amount of citric acid soluble which is expected to be available to plants easily.

Reagents

Citric acid (2%), other reagents as detailed earlier.

Procedure

To an accurately weighed quantity of the fertilizer (5 gm) in a litre bottle, add 500 ml of citric acid solution (10 gm of citric acid dissolved in 500 ml and adjusted to 20°C), shake the bottle for 30 minutes. At the end of the time, transfer this immediately to a fine filter and collect the filtrate.

Transfer an aliquot to a 500 ml stoppered conical flask marked at 150 ml. Dilute the sample to 100 ml and if the sample does not contain calcium, add 100 or 200 mg calcium carbonate and 5 N sodium hydroxide drop wise until a faint permanent turbidity or precipitate is formed. Dissolve the precipitate by the addition of dilute hydrochloric acid avoiding excess. Dilute to 150 ml and add 50 ml of citric - molybdic acid reagent and heat the solution to incipient boiling, maintaining it at that temperature for 3 minutes and then to boiling.

To this add 25 ml of quinoline solution drop wise by constant swirling initially and the rest in slow stream and keep the solution gently boiling during the addition. Immerse the flask in boiling water for 5 minutes, cool it to 15°C in running water.

Filter with suction the contents of the flask on a filter pulp pad and wash the flask and precipitate and filter with successive small washes of cold water until they are free from acid. Transfer the filter paper and the precipitate to the original flask, rinse the funnel with water and collect the rinsings in the flask. If necessary, add water to make up to 100 ml, stopper and shake it vigourously until the pulp and precipitate are completely dispersed. Then add a known volume of 0.5 N sodium hydroxide sufficient to dissolve the precipitate and be in excess. Shake the flask vigourously until all the precipitate dissolves. (If necessary to increase the dispersal of the precipitate, add a few drops of surface active reagent after addition of 0.5 N sodium hydroxide). Then add 0.5 - 1.0 ml of the indicator, titrate the excess of sodium hydroxide with 0.5 N hydrochloric acid until the indicator colour changes from violet to greenish blue and then very sharply to yellow at the end point. From this calculate the amount of 0.5 N sodium hydroxide actually consumed and then calculate the amount of citric acid soluble P_2O_5.

Calculations

Weight of sample taken = W gm

Volume of citric acid extract = 500 ml

Volume of citric acid extract taken = V ml

Volume of 0.5 N sodium hydroxide added = X ml

Volume of 0.5 N hydrochloric acid used for back titration=Yml

Volume of 0.5 N sodium hydroxide actually used = (X - Y) ml.

1 ml of 0.5 sodium hydroxide = 1.366 mg P_2O_5=1.366 /1000 gm

(X -Y) ml of 0.5 N sodium hydroxide = $\frac{1.366 \times (X-Y)}{1000}$ gm P_2O_5

This is found in "V" ml of citric acid extract.

Hence in 500 ml of total citric acid extract (=W gm)

$$= \frac{1.366 \times (X-Y)}{1000} \times \frac{500}{V} \text{ gm of } P_2O_5$$

Per cent of citric acid soluble P_2O_5

$$= \frac{1.366 \times (X-Y)}{1000} \times \frac{500}{V} \times \frac{100}{W}$$

16. FREE PHOSPHORIC ACID (as P_2O_5)

Principle

The free phosphoric acid is extracted from the fertilizer by acetone in a soxhlet apparatus and estimated by titration with standard alkali.

Reagents

Sodium hydroxide (0.1 N), bromocresol green indicator acetone.

Procedure

Weigh accurately a known weight of the sample and transfer it to a thimble, keep it in the extractor of a soxhlet's apparatus and extract it with acetone for 3 hours. Cool and distill off the acetone and take up the residue with water and make up the volume with water to 250 ml. Pipette exactly 100 ml of the extract and titrate with 0.1 N sodium hydroxide using bromocresol green as indicator till the colour changes from yellow to blue.

Calculations

Weight of fertilizer sample = W gm

Volume of extract prepared = 250 ml

Volume of extract taken for analysis = 100 ml

Volume of 0.1 N sodium hydroxide used = V ml

Normality of standard alkali = 0.1 N

Free phosphoric acid as P_2O_5 per cent by weight

$$= \frac{17.75 \times V \times \text{Normality}}{W}$$

C. POTASSIC FERTILIZERS

17. TOTAL POTASSIUM IN FERTILIZERS

Principle

The potassium present in the fertilizer is converted into potassium cobaltinitrite by reacting with sodium cobaltinitrite and estimated volumetrically using potassium permanganate.

Reagents

Sodium nitrite (35%), cobaltinitrite (20%), saturated sodium chloride solution, alcohol (35%), potassium permanganate (0.1 N), oxalic acid (0.1 N), glacial acetic acid.

Procedure

Weigh accurately a known amount of the potassic fertilizer (approximately 1 gm) and dissolve in a small amount of water and transfer to a 250 ml volumetric flask and make up to the volume. Then pipette 50 ml of the solution into a porcelain basin and evaporate to dryness on a water bath, till it is completely dry. Than keep it on a sand bath and roast it in a flame till grayish ash is obtained.

Extract the ash with hot water and filter through Whatman No. 40 filter paper and collect the clear extract in a beaker and evaporate to dryness on a water bath. Cool it and add 1.5 ml of glacial acetic acid followed by 10 ml saturated sodium chloride, stir the contents with a glass rod for 3 to 4 minutes and add 5 ml of sodium nitrite solution. Stir again for 2 - 3 minutes and allow it to react for about 5 minutes. Then add 5 ml of 20% cobaltinitrite solution rapidly and stir the contents, cover the beaker with a watch glass and keep it overnight under a bell jar.

Next day filter the precipitate through a gooch crucible fitted with asbestos, by transferring the precipitate without loss and washing 4 times with 8 to 10 ml of 35% alcohol. Finally wash the crucible with three small amounts of cold water to remove the alcohol.

To a 250 ml beaker, add a known amount of 0.1 N potassium permanganate, 150 ml of water and 5 ml of concentrated sulphuric acid. Transfer the precipitate from the crucible to the beaker and gently warm it. If the permanganate colour is discharged, add additional amounts of the permanganate and note the total volume added. Heat to boiling, then remove it and add an excess of 0.1 N oxalic acid. Warm till all the oxides of Mn have been dissolved and titrate the excess oxalic acid with standard permanganate. The difference between the amounts of permanganate and oxalic acid corresponds to the amount of 0.1 N potassium permanganate reduced by the sodium potassium cobaltinitrite precipitate. From this calculate the percentage of potassium present.

Calculations

Weight of fertilizer = W gm

Volume of solution prepared = 250 ml

Volume of aliquot taken = V ml

Volume of 0.1 N permanganate used = X ml

Volume of 0.1 N oxalic acid used = Y ml

Volume of 0.1 N permanganate actually reduced by the precipitate = (X -Y) ml

1 ml of 0.1 N permanganate = 0.00083 gm of K_2O

(X-Y) ml of 0.1 N permanganate = 0.00083 x (X-Y) gm of K_2O

This is present in V ml of the extract

250 ml of the extract (= W gm of fertilizer)

$$= 0.00083 \times (X-Y) \times \frac{250}{V}$$

Per cent of K_2O present in the fertilizer

$$= 0.00083 \times (X-Y) \times \frac{250}{V} \times \frac{100}{W}$$

If the potassium content obtained as K_2O has to be expressed in terms of K, then multiply the K_2O% by 0.83

18. POTASSIUM CONTENT IN MANURES

For estimation of potassium in manures like farm yard manure, compost or groundnut cake, prepare triple acid extract as given below and estimate the potassium using flame photometer.

Reagents

Triple acid

Procedure

Prepare the triple acid extract as indicated in the total phosphorus analysis of manures. Pipette out a known amount of triple acid extract, neutralise it with ammonium hydroxide and dilute it to known volume. After mixing the solution, measure the potassium content using the flame photometer and calculate the per cent of K in the manure.

Calculations

Weight of manure sample taken = W gm

Volume of triple acid extract prepared = 100 ml

Volume of triple acid extract taken = V ml

Volume of extract made up after neutralisation = 25 ml

Concentration of K as obtained from the standard curve=X ppm

$$\text{Per cent of K in the manure} = X \times \frac{100}{V} \times \frac{100}{W} \times \frac{25}{1{,}000{,}000}$$

Per cent of K_2O in the manure = K% x 1.2046

Precaution

Invariably the standard curve is prepared with ppm of potassium expressed in "K" and many a time in the calculations this is taken as "K_2O" and expressed as K_2O. This should be avoided. Vice versa may also happen and that too must be avoided.

ii. PERCHLORIC ACID METHOD

Principle

This method is recommended by the Fertilizer control order for adoption in the laboratories. It depends on the insolubility of potassium perchlorate and solubility of sodium perchlorate in alcohol. It is applicable in the presence of alkali metals, chlorides and nitrates but sulphates and ammonium salts must be absent on account of low solubility of sodium sulphate in alcohol. Phosphate if any, must be removed.

Reagents

Concentrated hydrochloric acid, barium chloride solution (10%), dilute hydrochloric acid, calcium oxide (finely ground), ammonia (sp.gr. 0.88), saturated aqueous solution of ammonium carbonate, perchloric acid (20%), methylated alcohol (95-96%), wash solution.

Procedure (for potassium salts-free of sulphate)

Dissolve an accurately weighed quantity of the fertilizer in water and transfer it to a 500 ml volumetric flask, make up to volume and allow it to settle.

Estimation

To a known volume of the extract in a basin, add 7 ml of perchloric acid, and place the basin on a hot plate or sand bath and evaporate the contents until white fumes are evolved. Cool the solution and dissolve the precipitate in a little hot water, add 1 ml of perchloric acid and concentrate to fuming. Then cool the residue, add 20 ml of alcohol and stir well. Allow it to settle and then pour the clear liquid through a dry filter paper. Then add 20 ml of alcohol to the basin, stir well and filter after cooling. Redissolve the precipitate on the paper and basin with the hot water. Add 2 ml of perchloric acid solution to the combined solution and evaporate to fuming stage. Cool the residue in the basin and thoroughly stir the contents with 20 ml alcohol. Allow the precipitate to cool, settle and then pour the clear liquid through a weighed gooch or sintered glass crucible, draining the precipitate as completely as possible, before adding 5 ml of wash solution. Wash the precipitate by decantation with small amounts of wash solution until free of acid. Then dry the precipitate at 100° C and weigh. The precipitate remaining is potassium perchlorate ($KClO_4$) and from this calculate the per cent of K_2O.

Calculations

Weight of sample taken = W gm

Volume of extract made = 500 ml

Volume of extract taken = V ml

Weight of precipitate = W_1 gm

This is from V ml of the extract.

Hence in 500 ml of the extract = $W_1 \text{ x } \frac{500}{V}$

Per cent of potassium perchlorate = $W_1 \times \frac{500}{V} \times \frac{100}{W}$

Per cent of K_2O present = $W_1 \times \frac{500}{V} \times \frac{100}{W} \times 0.34$

Modifications of the method

i. *If the fertilizer contains sulphates*: Dissolve an accurately weighed quantity of the sample in 300 ml of water and 20 ml of concentrated hydrochloric acid and heat the solution to boiling. To the boiling solution, add carefully drop by drop barium chloride solution till it is slightly in excess than that needed for precipitating all sulphates, cool it, transfer it to 500 ml volumetric flask, make up to mark, and filter through a dry filter. Take 50 ml of the filtrate and evaporate to dryness in a basin, moisten the residue with concentrated hydrochloric acid and evaporate to dryness, dissolve the residue with 5 to 10 ml of dilute hydrochloric acid and filter if necessary. Determine the potassium content as in the estimation of K.

ii. *If the fertilizer contains interfering ions like phosphates, iron, manganese or the fertilizer is a mixed fertilizer* : Weigh accurately about 10 gm of the sample and destroy the organic matter if any, by incinerating it at a temperature less than 500°C. Transfer the sample to a 500 ml beaker, add little water, concentrated hydrochloric acid and warm for 10 minutes. Dilute with water to about 300 ml and boil it. Then add about 10 gm of calcium oxide, make into a paste with water and boil it for 30 minutes with frequent stirring. Cool it, transfer it to 500 ml volumetric flask, make up to the volume and allow it to settle.

Then transfer 250 ml of the clear filtrate to another 500 ml volumetric flask, make it acidic and heat to boiling. Add carefully drop by drop, barium chloride solution until no further precipitation of sulphate occurs. Make the contents of the flask alkaline with ammonia and precipitate calcium and any excess of barium by adding ammonium carbonate solution till precipitation is complete. Then add 1 ml of ammonium oxalate solution, cool it and make up to 500 ml volume. Then filter through a dry filter paper or if the solution is clear, pipette directly 100 ml of the extract and evaporate to dryness in a basin. Expel all the ammonium salts from the residue by gentle heating over a low flame. Cool the residue, moisten with concentrated hydrochloric acid and again evaporate to dryness. Take up the residue with water and filter if necessary and determine the potash content as in the section on "estimation of K".

19. SODIUM TETRAPHENYLBORON METHOD

Principle

This is a direct volumetric method wherein the potassium is estimated using sodium tetraphenylboron and clayton chloride as indicator.

Reagents

Sodium hydroxide (20%), formaldehyde (37%), sodium tetraphenylboron (STPB) solution (approximately 12%), quaternary ammonium chloride solution (0.625%), clayton yellow indicator (0.04%), ammonium oxalate (4%), zephiran chloride, diglycol sterate.

Procedure

Transfer accurately weighed quantity of the fertilizer (2.5 gm) to a 250 ml volumetric flask and add 125 ml water, 50 ml saturated ammonium oxalate, 1 ml of diglycol sterate solution and boil for 30 minutes. Then add slight excess of ammonia and make it up to mark and filter through a dry filter. (In case of potassium salts without interfering substances of ammonia, calcium, aluminium etc., dissolve the fertilizer directly in water without addition of ammonia or ammonium oxalate).

Transfer 15 ml of the extract to 100 ml volumetric flask, add 2 ml of 20% sodium hydroxide, 5 ml of formaldehyde, and standard sodium tetraphenyl boron solution at the rate of 1 ml per per cent of K_2O expected plus 8 ml of sodium tetraphenyl boron to ensure complete precipitation. Dilute to volume with water, mix thoroughly and allow it to settle. Filter it through dry filter paper Whatman No. 12 or equivalent.

Transfer 50 ml filtrate to 125 ml conical flask, add 6-8 drops of clayton indicator and titrate the excess reagent with the standard zephiran solution.

Calculations

Weight of sample = 2.5 gm

Volume made up = 250 ml

Volume of extract taken = 15 ml

Percentage of K_2O per ml of STPB reagent
= (43 ml - ml zephiran) = 34.61

Percentage of K = 28.73/(43 ml - ml of zephiran)

ESTIMATION OF OTHER NUTRIENTS

20. ZINC IN ZINC SULPHATE

Principle

The zinc present in zinc sulphate is estimated by the EDTA method using erichrome black indicator.

Reagents

Disodium methylene diamine tetraacetate (EDTA) solution, (0.1 N) standard zinc solution, hydrochloric acid (1:1), ammonium hydroxide (20%), erichrome black T indicator, ammonium chloride, sodium cyanide, formaldehyde solution.

Procedure

Dissolve 0.5 gm of the material in 25 ml of water and add 0.1 gm of ammonium chloride and 0.5 gm of sodium cyanide. Add more cyanide if the precipitate formed does not dissolve completely. Add approximately 75 ml of water, 10 to 20 ml ammonia and sufficient amount of erichrome black T indicator to produce a red colour (only a few specks are required) and stir until it is dissolved. Add 20 ml of formaldehyde and titrate with EDTA solution to a clear blue end point. The titration should be carried out immediately after the addition of formaldehyde.

Calculations

Weight of the substance taken = W gm

Volume of EDTA used in blank = V_1 ml

Volume of EDTA used with the material = V_2 ml

$$\text{Amount of zinc (as Zn) in per cent} = \frac{(V_1 - V_2) \times 100}{W}$$

21. COPPER CONTENT

Principle

Sodium diethyldithiocarbamate reacts with slightly acid or ammonical solution of copper in low concentrations to produce a brown colloidal suspension of cupric diethyldithiocarbamate. The suspension can be extracted with an organic solvent and the extracted colour is determined spectrophotometrically.

Reagents

Ammonium citrate solution, sodium diethyldithiocarbamate (0.1%), standard copper solution, sodium sulphide, concentrated sulphuric acid, ammonium hydroxide, carbon tetrachloride, dilute nitric acid.

Special equipment

Spectrophotometer

Procedure

Weigh accurately about 5 gm of the sample and dissolve it in a small quantity of distilled water and 1 ml of concentrated sulphuric acid. Filter the solution into a 250 ml volumetric flask and make up to mark with water. Take 50 ml of the above solution in a beaker, heat, pass hydrogen sulphide gas or add sodium sulphide solution and ensure complete precipitation. Filter hot and keep the filtrate for other analysis.

Boil the residue with dilute nitric acid and filter if necessary. To the filtered solution, add concentrated sulphuric acid, evaporate it and dilute it with water. Filter and collect the filtrate for copper determination using diethyldithiocarbamate.

Transfer the filtrate to 100 ml separating funnel, add 2 ml of ammonium citrate solution and adjust the pH to 8.5 with ammonium hydroxide. The presence of ammonium ions helps to deionise the iron if present and to prevent rapid fading of the developed colour. Add 10 ml of sodium diethyl dithiocarbamate solution and develop the colour. Extract the colour complex with 5-10 ml of carbon tetrachloride and separate the carbon tetrachloride layer from the aqueous layer and centrifuge for 5 minutes to separate the water droplets. Transfer this carbon tetrachloride solution to a photometric cell and measure the absorbance at 440 mμ relative to the reagent blank. Calculate the correlated absorbance by subtracting the reading obtained from the solution containing no copper. The amount of copper in the extract can be found out from the standard curve.

Calculations

$$\text{Copper as Cu per cent} = \frac{W_1}{W_2} \times 100$$

Where W_1 = Weight of copper as determined in the given aliquot of the test solution

W_2 = Weight of material present in aliquot of the test solution

22. MAGNESIUM CONTENT IN FERTILIZERS

Principle

After removal of the copper present, the amount of magnesium is estimated by the EDTA method.

Reagents

Erichrome black T indicator, ammonium hydroxide - ammonium chloride buffer, standard magnesium solution, EDTA (0.1 N), dilute sulphuric acid, ammonium hydroxide, diammonium hydrogen phosphate.

Procedure

Take the filtrate from the copper estimation after removal of copper, add a few drops of concentrated nitric acid, boil, cool and then add 2 gm of solid ammonium chloride, boil and cool. Add ammonia till strong smell of ammonia is felt. Filter the precipitate through sintered crucible, take the filtrate, add dilute sulphuric acid till the solution is acidic using methyl red indicator. Heat it to boil and add excess of diammonium hydrogen phosphate with stirring. Add 10% ammonia till the solution is alkaline. White precipitate of zinc ammonium phosphate will be formed (The optimum pH for precipitation is 6.7. Allow it to stand for 3 hours).

Filter through Whatman filter paper No. 40 into a volumetric flask (100 ml) and make up the volume. Take 10 ml of the solution, add 20 ml of water, one ml of erichrome black T indicator and 20 ml ammonium hydroxide - ammonium chloride buffer solution. Heat it to 40 - 50°C, titrate with standard EDTA such that the colour changes from wine red to blue.

Calculations

Weight of material taken = W gm

Volume of extract made = 100 ml

Aliquot of solution taken = 10 ml

Volume of 0.1 N EDTA used = X ml

1 ml of EDTA equivalent = 0.2432 mg of Mg.

X ml of EDTA = 0.2432 x X mg of Mg

$$\text{Percentage of magnesium} = \frac{0.2432 \times X \times 10}{W \times 1000}$$

E. SAMPLING OF FERTILIZERS

Just as in any other materials, proper sampling of fertilizer ensures not only better precision in analysis but also helps to overcome number of legal aspects. As fertilizers are generally sold in bags, initially fertilizer bags truly representing all the other bags have to be selected. For this, random method of sampling (or identifying) the bags have been proposed to avoid any bias in selection.

First arrange all the bags in a systematic manner and count the total number of bags available. Depending upon the total number of bags decide on the number of bags to be selected for sampling using the table given below:

Total number of bags	Number of bags to be selected (n)	Total number of bags	Number of bags to be selected (n)
<10	1	600 - < 800	6
10 - <100	2	800 - <1000	7
100 - <200	3	1000 - <1300	8
200 - <400	4	1300 - <1600	9
400 - <600	5	1600 - <2000	10

Then start counting anywhere (any location) and count (N/n) = r number of bags, "r" being equal to the integer. Select every "r"th bag and draw them out for taking representing composite sample.

From each bag draw a suitable sample using a suitable sampling probe comprising of a slotted tube with solid cone tip made up of stainless steel or brass of length approximately 60-65 cm and diameter of the tube being approximately 1.5 cm, the slot being 1.2 to 1.3 cm. Pool all the collected samples and make it a composite sample by mixing properly.

While sampling the fertilizers the following precautions may be observed :

i. Samples should not be taken at a place exposed to rain or sun.

ii. Sampling instruments should be clean and dry

iii. The materials being sampled should be free of any other contamination

iv. The contents should be mixed as thoroughly as possible.

v. After preparing the composite sample, the sample should be stored in a clean, dry and air tight bottle (glass or polythene) or in thick gauge polythene bags and opened only just before the analysis.

vi. If they are intended for analysis in a fertilizer control laboratory the current instructions of sampling, packing and despatch should be strictly adhered to avoid any legal problems.

SAMPLING OF MANURES

In case of manures, if they are in heaps, then draw the samples either with a shovel or any other implement from different locations and at different depths. Select the places at random. Collect the samples from as many places as possible in the heap and inside the heap, pool them and mix them thoroughly and prepare the sample by quartering as described below:

Spread all the materials uniformly on a dry paper or cloth or dry clean floor. Divide it into four equal quarters and choose any diagonally opposite two, discarding the other two quarters. Then mix the materials from the chosen quarters together thoroughly and again divide it into four equal quarters. Select any two diagonally opposite two quarters, discarding the other two. Repeat the process till the bulk of the materials collected from the manure heaps are reduced to the required size (volume of 1 - 2 kg).

As the manures contain certain amount of moisture, dry them in the shade till it is completely dry (never dry the manures under the sun) and then transfer it to the container. During drying care should be taken not to permit any contamination from other sources or loss due to removal by air (wind) or volatilisation due to high temperature. If necessary powder the materials gently with wooden mallet before transferring it to the sample bottle.

Another important precaution to be observed is that as the manures are made up of different sized materials and also vary in their density, there is always a possibility of the particles segregating out and only particles of a particular size fraction or density getting sampled even during the quartering process. Hence thorough mixing and unbiased good sampling should be there while quartering as well as during the sampling of the materials for analysis.

As the moisture content in different samples vary widely, before estimating the nutrient content, first the moisture content should be estimated and then nutrient content expressed on moisture free basis. If the moisture percentage of the material is "M", then the nutrient content obtained in the estimation should be multiplied by 100/(100-M) to get the nutrient content on moisture free basis.

CHAPTER - 11

ANALYSIS OF PLANT MATERIALS

In the water and fertilizer samples, considerable amount of uniformity in composition could be expected within a batch or a sample bag. Hence sampling them is considerably simple one. But in the case of plants and plant materials, their composition vary from part to part of the same plant, one stage of growth to another stage of growth, even from time to time on the same day and also depending upon the soil conditions. Hence accuracy of analysis and later proper interpretation of the analysed data depends upon the collection of proper representative sample of the plant material for analysis. Hence great care has to be exercised in sampling of the plant materials.

First the selection of plants in a field should be done by random sampling. For this, select the required plant at random using either random numbers table or drawing lots or moving in a regular and at the same time in a zig-zag manner in the field or plot and marking the plants to be selected. Avoid atleast 2 rows on all sides of the plots and do not collect samples from those border rows as they are likely to exhibit the effects of nearby plots, adjoining bunds, excess moisture, shade effects etc. If the samples have to be drawn at periodical intervals, then draw the samples almost at the same time (hour) of the day to avoid any effect due to diurnal variations.

After collecting the plants, separate the various parts of the plant carefully and pool them up part wise and dry them in the shade first and then in the air oven. Moist or succulent plant materials will be difficult to grind in a mill due to lumping of materials. Hence these may be dried in shade and then in air oven. Drying at lower temperatures preferably between 60 and 70°C is preferred as at high temperatures there may be loss of nitrogen due to volatilisation.

After the material has dried completely, it may be ground in a mortor by hand or in a Wiley Mill using 200 mesh sieve. In case micronutrient content in the samples have to be estimated, then the plant materials may be ground in a stainless steel mortor or in a mill fitted with stainless steel blades to minimise the contamination in the samples with the micronutrients from metallic grinders. The powdered materials should be thoroughly mixed and stored in an air tight container till it is taken up for analysis.

The plant materials are generally analysed for their moisture content, ash, nitrogen, phosphorus, potassium, calcium, magnesium, micronutrients and crude fibre. The methods of analysis are basically the same as that adopted for analysing either water or fertilizers. The difference in the analysis is only in the digestion of the sample and preparation of the extract of the materials. Even this is similar in many respects to the preparation of the extracts of mixed fertilizers like the preparatioh of triple acid extracts of manures.

ANALYSIS OF PLANT MATERIALS

1. MOISTURE CONTENT

Principle

Samples of different plant materials even when stored in the same type of containers and under the same atmospheric conditions absorb different amounts of moisture. No two plant samples will have the same moisture content. Hence the composition of two plant samples can be compared only on moisture free basis. The moisture content is determined by heating a known weight of plant materials in an electric oven at 100 - 105°C to constant weight and then finding out the loss in weight due to drying.

Apparatus

Weighing bottle, air oven.

Procedure

Place the empty weighing bottle with the lid in an electric air oven maintained at 100 - 105°C for atleast 15 minutes. Then remove the weighing bottle and the lid to a desiccator, cool it and record the weight as quickly as possible in a balance.

Sample the plant material carefully and transfer 5-6 gm of the sample into the previously weighed weighing bottle and close the lid and find out total weight (weighing bottle + lid + plant material) accurately. Then keep the weighing bottle with the contents in the air oven at 100 - 105°C for 6 hours. Keep the lid open. In case different weighing bottles are kept in the oven, care has to be taken to see that the lids are not interchanged. For this purpose, it is better to label both the weighing bottle as well as the lid with the same number and this will prevent mixing up of containers. After drying, close the weighing bottle and transfer it to a desiccator, cool it and find out the weight. Again keep it in the oven for 30 minutes and find out its weight. Repeat it till constant weight is obtained. Calculate the loss in weight and express its percentage.

Calculations

Weight of empty weighing bottle = W_1

Weight of weighing bottle+plant material (before drying)=W_2

Weight of plant material = $(W_2 - W_1)$ gm

Weight of plant material + Weighing bottle (after drying) =W_3 gm

Weight of moisture removed = $(W_3 - W_2)$ gm

$$\text{Per cent of moisture} = \frac{(W_2 - W_3) \times 100}{(W_2 - W_1)}$$

2. ASH CONTENT

Principle

The ash content of the plant material gives a rough indication of the amount of inorganic constituents present in the plant sample. For this the organic matter is destroyed by heating and the material is converted into ash. The amount of ash is estimated by weighing.

Apparatus

Silica basin or crucible, muffle furnace, weighing bottle

Procedure

Heat the silica basin or crucible over a rosehead flame for about 15 minutes, then cool it in a desiccator and find out its weight. Transfer about 2 to 5 gm of the plant materials into the previously weighed silica basin or crucible and ignite it first slowly on a bunsen burner at low flame and after the substance chars, transfer it to the muffle furnace and heat it strongly at dull red heat (500 - 550°C) till a white ash is obtained. See that there is no carbon particles adhering to the sides. Then cool it in a desiccator and weigh it accurately. From this estimate the ash content.

Calculations

Weight of empty silica basin = W_1 gm

Weight of silica basin + plant sample = W_2 gm

Weight of plant sample = $(W_2 - W_1)$ gm

Weight of silica basin + ash = W_3

Weight of ash = $(W_3 - W_1)$ gm

$$\text{Per cent of ash} = \frac{(W_3 - W_1) \times 100}{(W_2 - W_1)}$$

Per cent of ash on moisture free basis

$$= \frac{(W_3 - W_1)}{(W_2 - W_1)} \times \frac{100 \times 100}{(100 - M)}$$

(The ash can be digested with concentrated hydrochloric acid and used for estimation of other nutrients like phosphorus, potassium, calcium and magnesium).

3. TOTAL NITROGEN (Macrokjeldahl method)

Principle

The nitrogen compounds in the plant materials are reduced by the salicylic acid to ammonia which is converted into ammonium sulphate by digestion with concentrated sulphuric acid. The digested material containing ammonium sulphate is distilled with excess alkali and the distillate is collected in 2%

boric acid or in 0.1 N sulphuric acid, using double indicator. From this the amount of nitrogen present can be calculated and by multiplying the nitrogen content by a suitable factor (5.8 to 6.35, usually 6.25) the crude protein content of the plant material is calculated. (For equations and reactions taking place, please refer to the section on method for total nitrogen estimation in manures).

Reagents

Same as in total nitrogen estimation in mixed fertilizers, boric acid (2%), double indicator (bromocresol green and methyl red indicator)

Procedure

Weigh accurately 0.5 to 1.0 gm of the plant material and analyse the same as per the procedure given for total nitrogen in manures. The evolved ammonia can be absorbed in 0.1 N sulphuric acid with methyl red indicator or in 2% boric acid containing the double indicator.

Calculations

i. In case 0.1 N sulphuric acid is used for absorption of evolved ammonia, follow the calculations as given in the analysis of manures for total nitrogen.

ii. In case ammonia is absorbed in "2% boric acid" then the following calculations may be adopted :

Weight of plant materials = W gm

Volume of 0.1 N sulphuric acid consumed for titrating the evolved ammonia = X ml

1 ml of 0.1 N sulphuric acid = 0.0014 gm N

X ml of 0.1 N sulphuric acid = 0.0014 x X gm N

This is present in "W" gm of the plant material

$$\text{Per cent of N in the sample} = 0.0014 \times X \times \frac{100}{W}$$

Moisture percentage in the sample = M

Per cent of nitrogen on moisture free basis

$$= 0.0014 \times X \times \frac{100}{W} \times \frac{100}{(100 - M)}$$

4. TRIPLE ACID EXTRACT OF PLANT MATERIALS (Wet digestion technique)

The triple acid extract of the plant materials is prepared for the estimation of total phosphorus, potassium, calcium and magnesium. The method is simple and less time consuming than the hydrochloric acid digestion. However greatest care has to be exercised in handling the acids, particularly the perchloric acid as it is highly dangerous.

Reagents

Triple acid (9 ml of concentrated nitric acid + 2 ml of concentrated sulphuric acid + 1 ml of perchloric acid)

Procedure

Weigh accurately a known amount of the plant material into a 100 ml conical flask and add 10 ml of triple acid mixture and cover the mouth of the flask with a small funnel. Digest the contents of the flask over a sand bath till a clear solution is obtained. Dilute with distilled water carefully and transfer it to a 100 ml volumetric flask and add the washings to the volumetric flask and make up to the mark and use the same for the following analysis.

5. PHOSPHORUS

i. Volumetric method

Principle

The phosphorus in the triple acid is precipitated as ammonium phosphomolybdate in nitric acid medium by the addition of ammonium molybdate. The precipitate is filtered and washed free of nitric acid and then dissolved in a known excess of alkali. The excess alkali is determined by back titration with standard acid using phenolphthalein as indicator as in the estimation of total phosphorus in farm yard manure.

$$6\ Na_2HPO_4 + 12\ (NH_4)\ MoOH + 23\ HNO_3$$

$$\text{------>}\ (NH_4)_3\ PO_4.12\ MoO_3 + 21NH_4\ NO_3 + 2\ NaNO_3 + 12\ H_2O$$

$$2\ \{(NH_4)_3\ PO_4\ .12\ MoO_3\} + 46\ KOH$$

$$\text{------>}\ 23\ K_2\ MnO_4 + (NH_4)_2\ MoO_4 + 2\ (NH_4)_2HPO_4 + 22\ H_2O$$

Reagents

Same as in the estimation of total phosphorus in manures.

Procedure

Pipette out 25 ml of triple acid extract into 250 ml beaker and proceed as in the estimation of total phosphorus in farm yard manure.

Calculations

Same as in the method for total phosphorus in farm yard manure.

ii. Colorimetric method

Just as in the case of estimation of total phosphorus in farm yard manure, estimate the total phosphorus in the triple acid using Barten's reagent and read the colour at 470 mμ in a spectrophotometer.

6. TOTAL POTASSIUM

Principle

The total potassium in the triple acid extract is neutralised with ammonia and then the solution is atomised in the flame. The flame excited atoms of potassium makes them emit radiations of specific wavelengths. The amount of radiation emitted is proportional to the concentration of potassium in the solution and is measured in a flame photometer with suitable filter that absorbs only the wavelength relating to potassium.

Special equipments

Flame photometer with K - filter.

Procedure

Pipette out 5 ml of triple acid extract and add 5 ml of ammonium hydroxide (1:4) just to neutralise the solution and feed this in the flame photometer after adjusting the same to zero with a blank and to 100 with 100 ppm K. Note the readings and find out the concentration of K using the standard curve. As indicated earlier, prepare a standard curve with solutions containing 20, 40, 60, 80 and 100 ppm.

Calculations

Weight of plant material = W gm

Volume of triple acid extract prepared = 100 ml

Volume of triple acid extract taken = 5 ml

Content of K in the plant material = X ppm

Moisture per cent = M

Per cent of total K in the sample on moisture free basis

$$= \frac{X}{10^6} \times 100 \times \frac{100}{W} \times \frac{100}{(100 - M)}$$

7. CALCIUM

i. Oxalate method

Principle

The calcium present in the sample could be estimated either by the oxalate method or by the versenate method. In the oxalate method, calcium is precipitated as calcium oxalate in the presence of ammonia by the addition of ammonium oxalate. The precipitate is washed free of chloride, then dissolved in dilute sulphuric acid and titrated against standard potassium permanganate. From the titre value, the amount of calcium present is calculated.

$CaCl_2 + (NH_4)_2C_2O_4$ -----> $CaC_2O_4 + 2\ NH_4\ Cl$

$2\ KMnO_4 + 3\ H_2SO_4$ -----> $K_2SO_4 + 2\ MnSO_4 + 3\ H_2O + 5\ (O)$ (i)

$5\ CaC_2O_4 + 5H_2SO_4$ ------> $5CaSO_4 + 5H_2C_2\ O_4$ (ii)

$5\ H_2C_2O_4 + 5\ (O)$ ----- $5\ H_2O + 10\ CO_2$ (iii)

Adding (i-iii) $2\ KMnO_4 + 8\ H_2\ SO_4 + 5\ CaC_2\ O_4$

---> $K_2\ SO_4 + 2\ MnSO_4 + 5\ CaSO_4 + 8\ H_2O + 10\ CO_2$

Reagents

Potassium permanganate (0.1 N), ammonium chloride, saturated ammonium oxalate, dilute sulphuric acid.

Procedure

Pipette out 25 ml of triple acid extract into a 250 ml beaker and add 20 ml of distilled water and 1 gm of solid ammonium chloride till the litmus paper turns blue. Heat it to boiling and then remove the flame. Add 10 ml of saturated ammonium oxalate solution slowly with constant stirring and boil for 2 to 3 minutes over a rose head flame. Then allow it to cool and allow the precipitate to settle. Test for the completion of precipitation by adding a small quantity of ammonium oxalate solution through the sides of the beaker. Keep the beaker over a sand bath, digest it for 30 minutes and filter through Whatman filter paper No. 40, washing the precipitate till free of chloride, with luke warm water (free of chloride may be tested with silver nitrate plus nitric acid). When the filtrate is free of chloride (free of oxalate) make a hole in the filter paper by means of glass rod and wash down the precipitate into the same beaker in which calcium was precipitated. Wash the filter paper first with hot water and finally with dilute sulphuric acid behind the folds also and collect the entire quantity in the beaker.

Add 10 ml of dilute sulphuric acid in excess to the beaker, heat it to 70°C (bearable warmth) and titrate it against 0.1 N potassium permanganate till a faint permanent pink colour persists. Transfer the filter paper from the funnel into the beaker and if the pink colour disappears continue the titration till the permanent pink colour persists. Do not pulp the filter paper as pulped filter paper will reduce more of permanganate and this will introduce a significant error.

Calculations

Weight of plant material taken = W gm

Volume of triple acid extract made up = 100 ml

Volume of extract taken for analysis = 25 ml

Volume of 0.1 N permanganate used in the titration = V ml

1 ml of 0.1 N permanganate = 0.002 gm Ca.

V ml of 0.1 N permanganate = 0.002 x V gm of Ca

This is present in 25 ml of extract

100 ml of triple acid extract (= W gm of plant material)

$$= 0.002 \times V \times \frac{100}{25}$$

Per cent of calcium (Ca) present in the sample

$$= 0.002 \times V \times \frac{100}{25} \times \frac{100}{W}$$

Per cent of calcium (Ca) present on moisture free basis

$$= 0.002 \times V \times \frac{100}{25} \times \frac{100}{W} \times \frac{100}{(100 - M)}$$

(Moisture = M%)

If the amount of calcium is required to be expressed in terms of CaO, then multiply the amount expressed as Ca by 1.399.

8. CALCIUM AND MAGNESIUM (EDTA method)

Principle

Calcium and magnesium when reacted with EDTA gets complexed in the following order; calcium and then magnesium as discussed in the section on water analysis.

In the case of plant materials, the amount of calcium and magnesium present is estimated by the EDTA method using the triple acid extract and following the same procedure as described under water analysis.

Reagents

EDTA (0.01 N), ammonium chloride - ammonium hydroxide buffer solution, erichrome black T indicator, murexide, sodium hydroxide (10%).

Procedure

For estimation of both calcium and magnesium, pipette out 10 ml of the triple acid extract, dilute it with 20 ml of distilled water and follow the procedure as outlined in the case of waters.

Calculations : (same as in the case water samples)

Weight of plant materials taken = W gm

Volume of triple acid extract prepared = 100 ml

Volume of extract taken for analysis = 10 ml

Volume of 0.01 N versenate used for calcium plus magnesium = X ml

Volume of 0.01 N versenate used for calcium estimation=Y ml

i. Calcium

1 ml of 0.01 N versenate = 0.0002 gm Ca

Y ml of 0.01 N versenate = 0.0002 x Y gm of Ca

This is present in 10 ml of extract.

100 ml of triple acid extract (= W gm plant sample) contains

$$= 0.0002 \times Y \times \frac{100}{10} \text{ gm of Ca}$$

$$\text{Per cent of calcium (Ca)} = 0.0002 \times Y \times \frac{100}{10} \times \frac{100}{W}$$

Per cent of calcium (Ca) on moisture free basis

$$= 0.0002 \times Y \times \frac{100}{10} \times \frac{100}{W} \times \frac{100}{(100 - M)}$$

(Moisture = M%)

ii. Magnesium

1 ml of 0.01 N versenate = 0.000243 gm Mg

(X - Y) ml of 0.01 N versenate = 0.000243 x (X-Y) gm of Mg

This is present in 10 ml of extract

100 of extract (= W gm of sample)

$$= 0.000243 \times (X - Y) \times \frac{100}{10} \text{ gm of Mg}$$

Per cent of magnesium (Mg)

$$= 0.000243 \times (X - Y) \times \frac{100}{10} \times \frac{100}{W}$$

Per cent of magnesium (Mg)

$$= 0.000243 \times (X - Y) \times \frac{100}{10} \times \frac{100}{W} \times \frac{100}{(100 - M)}$$

(Moisture = M%)

If the contents of calcium and magnesium have to be expressed in their oxide forms (CaO, MgO) the following factors may be used :

Multiply the amount of Ca by 1.399 to get the CaO content

Multiply the amount of Mg by 1.658 to get the MgO content

9. ETHER SOLUBLE EXTRACTS (CRUDE FAT)

Principle

For estimating the crude fat content, the plant material is extracted with petroleum ether boiling at 40-60°C using a Soxhlet apparatus and the amount of ether extractives estimated. This is reported as crude fat content of the plant material.

Reagents

Petroleum ether (40-60°C)

Special apparatus

Soxhlet apparatus

Procedure

Transfer an accurately weighed quantity of the plant material (about 3 to 5 gm) into the thimble and keep it in the extractor of the soxhlet apparatus. Weigh the extracting flask (bottom flask) and then add about 100-150 ml of petroleum ether and set the soxhlet apparatus. Warm the same over a sand bath. The warm ether will vapourise and condense in the upper extractor and extract the ether soluble materials. When the chamber is filled, it will get siphoned off. Allow atleast 8 siphonings to take place to ensure complete extraction of ether solubles. Then allow the ether in the last siphon to vapourise and collect in the upper chamber, till it is nearly full. Stop the evaporation just before the upper flask becomes full, remove the lower flask, evaporate the remaining ether by keeping it on a water bath. Cool it in a desiccator and then weigh. The difference in weights gives the amount of materials soluble in ether.

Calculations

Weight of plant materials taken = W gm

Weight of bottom flask + ether extractives = W_1 gm

Weight of the bottom flask = W_2 gm

Weight of ether extractives alone = $(W_1 - W_2)$ gm

$$\text{Per cent of ether extractives} = \frac{(W_1 - W_2)}{W} \times 100$$

Per cent of ether extractives on moisture free basis (moisture = M%) (Crude fat per cent)

$$= \frac{(W_1 - W_2)}{W} \times 100 \times \frac{100}{(100 - M)}$$

10. CRUDE FIBRE CONTENT

Principle

Crude fibre is the fraction of carbohydrates consisting of cellulose, hemicellulose, pentosans and lignins which are not digestible in dilute acids and alkalis. The acid-alkali insoluble material is separated and estimated. As inorganic constituent also form part of the residue, the residue is ignited and the weight of ash deducted from the residue. This gives the weight of crude fibre.

Reagents

Sulphuric acid (1.25%), sodium hydroxide (2%), phenolphthalein, potassium hydroxide (0.1 N), acetone.

Procedure

Transfer an accurately weighed quantity of the plant material into a 400 ml beaker, add 200 ml of 1.25% sulphuric acid and mark the level in the beaker. Boil the contents for exactly 30 minutes maintaining the volume at 200 ml by adding hot water. Filter through a muslin cloth and wash the residue with hot water till the filtrate runs free of acid as tested by collecting the filtrate to 1/2 to 3/4 of full of the test tube and adding one drop of phenolphthalein and a drop of 0.1 N potassium hydroxide. If a pink colour persists, it indicates that the filtrate is free of acid.

Transfer the residue completely to the same beaker, add 75 ml of hot water and 100 ml of 2.5% sodium hydroxide and fill up to the mark with hot water. Boil for exactly 30 minutes maintaining the volume as before, then filter it through the same muslin cloth and wash it with water till the filtrate is free of alkali (adopt the same test but the colour of the solution is colourless). Then wash the residue with 10 ml of acetone to remove any fat material and wash it with hot water 3 times to remove the acetone.

Transfer the residue to a previously weighed silica basin and evaporate all the water present by keeping it on a water bath and finally in air oven. Cool it in a desiccator and find out the weight of the residue. Then ignite the residue till it is completely ashed (white) and find out the weight of ash present. From this calculate the percentage of crude fibre.

Calculations

Weight of plant material taken = W gm

Weight of empty silica basin = W_1 gm

Weight of silica basin + residue = W_2 gm

Weight of silica basin + ash = W_3 gm

Weight of crude fibre = $W_2 - W_3$ gm

$$\text{Per cent crude fibre} = \frac{(W_2 - W_3)}{W} \times 100$$

Per cent of crude fibre on moisture free basis (Moisture = M%)

$$= \frac{(W_2 - W_3)}{W} \times 100 \times \frac{100}{(100 - M)}$$

CHAPTER - 12

ANALYSIS OF SOILS

As in the case of any analytical work, proper sampling is the basic requirement in soil analysis also for correct assessment of its properties. Materials like water and fertilizers are generally uniform and homogeneous in composition and hence collection of representative sample is comparatively simple. But sampling of soil is complicated due to the high heterogeneity and also as one kilogram of soil drawn for analysis is expected to truly represent more than 2 million kilograms of soil in the field if the area it represents is one hectare and depth about 15 cm. If the area is more, or the depth is more, then the mass it represents is still more. Hence a small error committed during sampling could vitiate the results significantly.

Soils are analysed to assess the physico-chemical properties with reference to the crop to be grown later. Hence it has to be sampled either few days before sowing the next crop or after the harvest of the previous crop. The following broad guidelines will help in proper sampling of soils:

a. Samples should be collected before the application of fertilizers.

b. Sampling of soils during rainy season should be avoided as far as possible.

c. Variations in slope, colour, texture, crop growth and management practices should be considered and separate samples should be collected for each one of the areas differing in any of the above aspects.

d. Also separate soil samples should be collected from areas suspected to be saline and/or alkaline.

e. Recently fertilized soils, areas near the furrows, drainage channels, old bunds, marshy spots, areas close to compost or manure pits, areas close to factories emitting different types of waste particles/pollutants etc., should not be sampled and pooled with other samples.

f. Depending upon the root pattern of crops, sampling depths have to be varied. In case of shallow rooted crops, soil samples from the surface (0 - 15 cm) may be collected while in the case of deep rooted crops, soil samples up to the maximum root depth of the crop must be collected, depth wise or horizon wise.

g. Plant materials, roots, undecomposed or decomposing organic matter visible to the naked eye and present on the surface must be removed before sampling.

h. Sampling tools should be clean so that the contamination from the tools can be avoided. In case samples have to be collected for taking up micronutrient analysis, sampling should be done preferably with stainless steel tool or tools made up of such materials that will not contaminate the samples.

SAMPLING TECHNIQUE

The area from which soil samples have to be collected should be surveyed and delineated based on their uniformity and also based on the guidelines detailed above. For each area, 10 to 15 soil samples of 1 to 2 kg weight may be collected for each depth/horizon separately from different locations in the apparently homogeneous field after scrapping the surface of the litter or undecomposed materials. In case of sampling of surface soils, a "V" cut may be made to the required depth with a spade or hand hoe. An augur may be used for collection of soil unless profile pits have already been opened for this purpose.

The volume of soil collected from each pit should as far as possible be same. The samples should be collected at random by following a zig-zag movement. All samples representing one homogeneous area may be pooled and mixed together and reduced to 1 to 2 kg size following the quartering method described below;

Quartering of soil samples : All samples representing one single area may be pooled on a clean sampling paper or on a hard clean floor, mixed thoroughly and divided into four quarters. Out of these, two opposites quarters may be discarded and the rest pooled and the process of quartering, discarding, quartering etc., repeated till the soil sample left with is around 1 to 2 kg in weight.

The sample thus collected should be dried in the shade at the room temperature and when dry, transferred to a plastic or clean cloth bag. The containers should be tightly closed and labelled with all information like sample number, place of sampling, depth of sampling, owner's name(s), previous crop, next crop to be grown, any peculiarity observed etc. The cloth or plastic bag used should be new and if it has not been used earlier, it may be checked for any contamination within the bag before filling up the same with the sample. If the sampling bag is old one, then those bags should not be used at all for soil samples. The labels on the soil samples should be legible and give as much information as possible. For writing the labels, ball point pens or copying pencils may be used. The collected soil samples should be protected against any possible contamination from rain, fertilizers, chemicals, manure and other contaminating materials. They may be stored in dry, well ventilated spots before analysis.

As soon as these soil samples are received in the laboratories, the information available may be recorded in a book and a fresh laboratory number assigned to it. Again the sample may be dried in the shade, crushed with a wooden mallet gently, then pooled and generally passed through 2 mm sieve unless otherwise indicated. In case of undisturbed soil - core samples collected for analysis, the core samples should not be crushed but they should be preserved as such. The other samples may be ground to pass through the desired mesh sieves and transferred to a glass or polythene containers with proper labels. In certain cases, where the percentage of gravel or other bigger sized particles are needed that information may also be recorded.

Sampling tool

Simple soil tube (augur), screw type augur, spade, handhoe etc., are commonly used for sampling. For sampling soft and moist soil, the tube augur, spade or handhoe may be sufficient while for hard or dry soil, a screw type augur may be used. A post hole augur will be useful for sampling excessively wet areas like rice fields. For sampling soils up to 15 cm depth, a spade or handhoe may be sufficient and for lower depths augurs may be used unless profile pits or open cuts are available.

A. PHYSICAL PROPERTIES

1. TEXTURE

Principle

Texture is defined as the fineness or coarseness of the soil and varies with the relative proportions of soil particles of varying sizes. Based on the size of the individual particles, they are classified as gravel, sand, silt and clay. As clay particles are very small in size, they are not visible to the naked eye while sand particles can be visually observed. The following are some of the characteristics used for classifying soils into different textural class.

Name of the particle	Diameter (mm)	Visibility	Feel by finger	Minerals
Coarse sand	2.00 - 0.20	Naked eye	Very gritty	Primary minerals
Fine sand	0.20 - 0.02	Naked eye	Gritty	Primary minerals
Silt	0.02 - 0.002	Ordinary microscope	Flour like	Primary and Secondary
Clay	< 0.002	Electron microscope	Very small	Mostly Secondary

When the soil contains considerable quantities of particles of at least two sizes, it is called a loam and may be further classified as clay loam, sandy loam etc., depending upon the size of the fractions present in larger quantities and also according to proportions of each with respect to other. If the soil consists mostly of particles of one size, it is classified based on the particular size that is dominating viz. clay, silt, sandy soil. The texture of soil is determined qualitatively by the field or feel method and quantitatively by the hydrometer or International pipette method.

1. Field method (Feel method)

This method provides an approximate estimation of the texture but in the hands of an experienced worker, the results by this method will be as good as any other laboratory method.

Procedure

Moisten a small quantity of dry soil with water and mix it thoroughly on a glass or porcelain dish so that a soft ball is formed. Then work it further until stiff, and squeeze it between the thumb and forefingers. Press the thumb forward gradually forming the soil into a ribbon. By observing the feel to the fingers, ease of forming a ball, stickiness or grittiness and whether the soil formed a ribbon or merely crumbles on squeezing, the soil can be classified. The important characters of certain classes are listed below:

a. If the soil turns into a ribbon easily and remains long and flexible upon pressing, the soil is probably clay or silty clay and is considered heavy soil or fine textured soil. Such soils are sticky and plastic.

b. If the soil ball forms a ribbon but breaks into 2 to 2.5 cm long pieces, the soil is moderately fine textured and may be either a clay loam or silty clay loam and is moderately sticky and plastic.

c. If it does not form into ribbons but breaks off into small bits (<2 cm), it is probably silty loam, sandy loam or loam. The soil is medium textured or moderately coarse textured.

d. If the soil feels smooth and talc like with no grittiness and silt predominating, the soil is medium textured. If it is slightly gritty, yet fairly smooth and talc-like, it is probably a loam or silt loam and hence a medium textured soil.

e. If the soil is extremely gritty without any smoothness, it is either a sandy or moderately coarse textured soil.

f. If it is made up of gritty material and leaves no stain on hand, it is a sandy soil or coarse textured soil.

Summarising the above, the various soil textural classes can be characterised as follows:

Soil class	Feel	Moist condition
Sand	Very gritty	Does not form ball or stain fingers
Loamy sand	Very gritty	Forms balls, broken easily, stains fingers slightly
Sandy loam	Moderately gritty	Forms firm balls, broken easily, stains fingers
Loam	Neither gritty nor smooth	Forms firm balls- does not ribbon-stains fingers
Silt loam	Smooth or sticky buttery feel	Forms firm balls - tending to ribbon with flakky surfaces - stains fingers
Clay loam	Moderately sticky- slightly gritty	Forms moderately hard balls when dry - ribbons on squeezing that easily breaks - stains fingers
Silty clay loam	Moderately sticky-very smooth	Forms moderately hard balls when dry - ribbons on squeezing that easily breaks - stains fingers
Clay	Sticky	Forms balls when dry - cannot be crushed by fingers - ribbon breaks to less than 2 cm easily - stains fingers

ii. Laboratory techniques

The mechanical composition of the soils is assessed in the laboratory by either of the following methods : a. Hydrometer method b. International pipette method. The hydrometer method is quick and fairly accurate for many purposes. The International method gives very accurate values of each of the soil separates and is useful for most of the fundamental studies but at the same time cumbersome and time consuming.

The principle in both these methods is that the density of suspension at a given depth decreases as an initially homogeneous dispersed suspension settles. The rate of decrease in density at any given depth is related to the settling velocities of the particles and obeys the Stoke's law which is

$$V = \frac{2gr^2\ (d_1 - d_2)}{g\ \eta}$$

where V = velocity of fall cm/sec.

r = radius of particles (cm)

d_1 = density of particles gm/ml

d_2 = density of liquid gm/ml

η = viscosity of the liquid (in poises)

g = acceleration due to gravity m/sec^2

a. HYDROMETER METHOD

Principle

The principle indicated above is utilised in estimating the amount of different sized particles present in the soil. Here the soil particles are dispersed using an electrical stirrer and dispersion agents and then estimated by use of hydrometer.

Reagents

Sodium hexametaphosphate (5% solution), hydrogen peroxide (30%), amyl alcohol.

Apparatus

Cylinder with rubber stopper, thermometer, Bouycous hydrometer, electrical stirrer with dispersing cup.

Procedure

Take 10 ml of sodium hexametaphosphate solution in one litre cylinder and make up to one litre with distilled water. Lower the hydrometer into the solution and note the reading and this gives the calibration correction (Rb).

Weigh two 40 gm of soil samples and dry one of them in an air oven at 105°C to determine the moisture content of the soil.

Transfer the other 40 gm of soil to a 600 ml beaker and add 200 ml of distilled water and 4 to 5 ml of hydrogen peroxide. Cover the beaker with watch glass and keep it on water bath till frothing is over, indicating the decomposition of organic matter. Cool the beaker and repeat the process till there is no frothing or the suspension does not change its colour.

Keep the beaker on water bath for 2 hours to remove hydrogen peroxide, then transfer the sample to the dispersing cup adding little quantity of water and 10 ml of sodium hexametaphosphate and keep it for 15 minutes. Stir the suspension for 10 minutes using an electrical stirrer. The time may have to be varied from 5 minutes for sandy soils to 30 minutes for heavy soils.

Transfer the contents to a one litre cylinder and make up its the volume with distilled water. Close the cylinder with a rubber stopper and shake the cylinder for a minute, keeping it in horizontal position. Place the cylinder on a flat table and start the stop watch. Insert the hydrometer carefully into the cylinder after 3 minutes 50 seconds and take the reading at exactly 4 minutes. By this time particles larger than 0.02 mm diameter would have settled below 10 cm depth. If the suspension is frothy, add a drop of amyl alcohol and take the readings. Record the temperature of the suspension. Keep the suspension undisturbed for 1 hour 59 minutes and 50 seconds and reinsert the hydrometer and take the readings at 2 hours. By this time, particles larger than 0.002 mm (sand and silt) would have settled down. (While inserting the hydrometer, there should be least possible disturbance to the suspension).

The hydrometer is calibrated at 67°F (20°C). In case the temperature of the suspension is different, a correction must be applied. For each degree above or below 65°F, it should be multiplied by 0.2

Hydrometer reading at 4 minutes = R_1

Hydrometer reading at 2 hours = R_2

Weight of soil (moisture free) = W gm

Temperature correction = r

Calibration correction = Rb

The percentage of dispersed material remaining in the suspension at the specified time is :

$$\text{Per cent of silt + clay} = \frac{R_1 - Rb + r}{W} \times 100$$

$$\text{Per cent of clay} = \frac{R_2 - Rb + r}{W} \times 100$$

Per cent of sand = 100 - (silt% + clay%)

The textural class is found out using the textural triangle.

b. INTERNATIONAL PIPETTE METHOD

Principle

In this estimation, the Stoke's principle is made use of for separation of different sized fractions.

Reagents

Hydrogen peroxide (6%), hydrochloric acid (0.2 N), sodium hydroxide (1 N)

Apparatus

Cylinder without spout, mechanical stirrer, stop watch

Procedure

Weigh exactly 20 gm of air dried soil and transfer it to a 500 ml beaker. Add 60 ml of 6% hydrogen peroxide solution to remove the organic matter (this is not necessary for soils with low organic matter) and 200 ml of 0.2 N hydrochloric acid. Stir well and keep it on water bath for 30 minutes until the effervescence ceases. Filter through Whatman No. 50 filter paper till the filtrate is free of chloride (test it with silver nitrate solution). Transfer the residue to another 500 ml beaker and dilute it to 400 ml of water. Add 8 ml of 1 N sodium hydroxide and stir it well with mechanical stirrer for 10 minutes (or use end over end shaker and shake it for 30 minutes). Transfer the suspension into 1000 ml measuring cylinder without spout and make up to mark (1000 ml) with distilled water, Cover the cylinder with rubber stopper and shake it thoroughly by repeated inversions.

Remove the rubber stopper and place the cylinder under a Robinson's pipette and start a stop watch. Do not disturb the suspension for 4 minutes and at the end of 4 minutes, lower the pipette to 10 cm depth and pipette out 20 ml of the suspension and deliver it into a porcelain dish, evaporate it on a water bath at 105°C and weigh it. The time of settling should be varied depending upon the temperature of suspension. This gives a measure of silt and clay.

Again shake the cylinder well and keep it for 6 hours and at the end of the period, pipette out 20 ml of suspension into another weighed porcelain dish, evaporate it on a water bath at 105°C and weigh it. This gives the amount of clay alone.

After taking out the sample for clay estimation, decant major portion of the suspension, transfer the remaining one to a tall form beaker of height greater than 12 cm, wash the sediment with water and allow it to stand for 4 minutes. Pour off the supernatent liquid. Repeat the process till the supernatent liquid is clear. Transfer the residue to a weighed porcelain basin, dry it at 105°C and this gives the amount of coarse and fine sand present in 20 gm soil.

The above residue (coarse sand plus fine sand) is transferred to a 0.2 mm sieve and sieved, the material passing through is collected and this gives the weight of fine sand or the material remaining on the sieve gives the amount of coarse sand. The particular fraction that is collected is transferred to a weighed porcelain dish, dried at 105°C and weighed and this gives the amount of that size fraction,

Calculations

Weight of soil taken = 20 gm

Weight of porcelain dish + clay + silt = W_1 gm

Weight of empty dish = W_2 gm

Weight of clay + silt in 20 ml suspensions = $W_1 - W_2$ gm

Weight of sodium hydroxide alone = 0.0064 gm

Weight of clay + silt = $(W_1 - W_2 - 0.0064)$ gm

$$\text{Per cent of clay + silt} = (W_1 - W_2 - 0.0064) \times \frac{1000}{20} \times \frac{100}{20}$$

Clay

Weight of dish + clay + sodium hydroxide = X_1 gm

Weight of empty dish = X_2 gm

Weight of sodium hydroxide = 0.0064 gm

Weight of clay alone = $(X_1 - X_2 - 0.0064)$ gm

Per cent of clay = $(X_1 - X_2 - 0.0064) \times \frac{1000}{20} \times \frac{100}{20}$

Per cent of silt = (per cent of clay + silt) - (per cent clay)

Coarse sand + fine sand

Weight of dish + coarse sand + fine sand = Y_1 gm

Weight of empty dish = Y_2 gm

Weight of coarse sand + fine sand = $(Y_1 - Y_2)$ gm

Per cent of fine sand + coarse sand = $(Y_1 - Y_2) \times \frac{100}{20}$

Coarse sand

Weight of dish + coarse sand = A_1 gm

Weight of empty dish = A_2 gm

Per cent of coarse sand = $(A_1 - A_2) \times \frac{100}{20}$

Per cent of fine sand = (Per cent coarse sand plus fine sand) – (per cent of coarse sand)

Based on the clay, silt, fine sand and coarse sand, the textural class is determined, using the textural class triangle.

2. BULK DENSITY

i. Core sampling technique

Principle

Bulk density of a soil indicates the degree of compactness of the soil and is defined as the mass per unit volume which includes space occupied by solids and pore space.

$$\text{Bulk density } D_b = \frac{\text{Mass (W)}}{\text{Volume (V)}} \text{ gm/ml}$$

Bulk density varies with particle size distribution, organic matter content, mechanical composition, depth of soil. Increase in content of clay or organic matter decreases soil bulk density and it increases with depth due to compactness. Bulk density is used for estimating porosity, void ratio, weight of furrow slice and for converting soil moisture from weight to volume basis. In this a core sample is drawn and from its weight and volume, the bulk density is estimated.

Apparatus

Core tube sampler, vernier calipers, moisture bottle, Keen's box.

Procedure

Measure the length and diameter of the core tube using vernier calipers. In the selected area, insert the core tube vertically into the soil by hammer. Remove the soil sticking to the outside of the core tube to enable easy withdrawal of the core tube along with the core from the field. Close the bottom of the core tube with hand or other means while withdrawing the tube to prevent the soil sliding from the core tube. After removing the excess soil, separate out the core sample and weigh the same.

Transfer 20-30 gm of the homogeneous soil sample to a moisture box and find out the weight of the box plus soil. Dry the same in an oven at 105°C for 24 hours and find out the oven dry weight. From this calculate the moisture and also dry weight of the core.

Calculations

Length of core tube = L cm

Diameter of the core tube = d cm

Volume of the core sample = $\pi \left(\frac{d}{2}\right)^2$ L ml = V ml

Weight of core sample = W_1 gm

Moisture per cent = M %

Weight of dry core sample = $W_1 \times \frac{100 - M}{100}$ = W gm

Bulk density = $\frac{W}{V}$

Precautions

i. Soil samples should not be collected in very wet or dry soil condition.

ii. The core sampler should have large diameter to reduce errors due to the compression during sampling. The diameter should be more than 5 cm.

ii. Wax Coating Method

Principle

In case where core samples cannot be collected, this method

can be used. Here a clod of the soil is collected, coated with wax to prevent it crumbling when put in water and then its weight and volume are estimated. From this the bulk density is calculated.

Procedure

Take a core or clod of soil in the field and find out its weight (W gm). Then tie the clod carefully with a small twine and immerse the clod in a melted wax and take it out and allow it to dry.

Take a known volume of water (150 ml) in a graduated measuring cylinder (250 ml) with a wide mouth and immerse the wax coated clod in the water contained in the cylinder. Note the increase in the volume of water in the cylinder and from that calculate the bulk density.

Calculations

Weight of clod = W gm

Initial volume of water in the cylinder = V_1 ml

Final volume of water in the cylinder after immersing the clod = V_2 ml

Volume of clod = $V_2 - V_1 = V$ ml

Bulk density of the soil = $\frac{W}{V}$ gm/ml

3. PARTICLE DENSITY OF SOIL

Principle

Particle density refers to the density of the soil solids and is defined as the mass per unit volume of soil solids. Since the volume is exclusive of pore space,

$$\text{Particle density} = D_p = \frac{\text{Mass (W)}}{\text{Volume of solids } (V_s)} \text{ gm/ml}$$

the particle density is higher than that of bulk density. Particle density is used in the estimation of porosity, void ratio and settling velocity of the soil particles using Stoke's law.

The particle density of most soils range between the narrow limits of 2.60 to 2.70 gm/ml with an average value of 2.65 gm/ml as most of the mineral particles are constituted of mainly quartz, feldspars and other silicates whose densities also vary within the same limits. If the soil contains high amounts of organic matter or heavy minerals such as magnetite, garnet, epidote etc., the bulk density may be less than the lower limit or more than the higher limit respectively. The particle density of soil does not vary due to the mechanical manipulations.

For determining particle density, the mass and volume of the soil solids have to be estimated. Mass is found by weighing the dry soil while volume of the soil particles is estimated by measuring the displacement of water by the soil particles. The accuracy of this method depends upon how closely the volume of the displaced water approached that of the true volumes of soil particles.

Apparatus

Pycnometer (specific gravity bottle)

Procedure

Transfer 10 gm of oven dried soil to a clean, previously weighed pycnometer and fill up the same with water up to half the volume of the pycnometer. Remove the entrapped air by gently boiling and then completely fill up the pycnometer with boiled and cooled distilled water. Insert the stopper and wipe out the pycnometer with filter paper if any water is sticking to the sides and find out its total weight (W_1). Empty the pycnometer and fill it up with boiled, cooled distilled water and find the weight (W_2) after closing the stopper and cleaning the outside of the pycnometer.

Calculations

Weight of oven dry soil = W gm

Weight of pycnometer with the soil + water = W_1 gm

Weight of pycnometer with water alone = W_2 gm

Volume of oven dry soil = $(W_2 + W - W_1) \times$ density of water

$$\text{Particle density} = \frac{\text{Weight of oven dry soil}}{\text{Volume of oven dry soil}}$$

$$= \frac{W}{(W_2 + W - W_1) \times \text{density of water}}$$

$$= \frac{W}{(W_2 + W - W_1)}$$

assuming the density of water as 1.

METHOD II

Procedure

Particle density can be determined using a measuring cylinder also. Transfer an accurately weighed quantity of soil (20 gm) to a measuring cylinder and after gently tapping the cylinder, note the volume, then add 50 ml of water slowly through the sides of the cylinder and allow it to soak completely. Then record the final volume of the soil plus water. Care should be taken to see that 5 ml of water is present on the surface at the end of the experiment.

Calculations

Weight of soil taken = W gm

Volume of soil = V ml

Volume of water added = V_1 ml

Total volume of soil + water (expected) = $(V + V_1)$ ml

Volume of soil + water at the end of the experiment = V_2 ml

$$\text{Particle density} = \frac{\text{Weight of soil}}{\text{Volume of soil}} = \frac{W}{V - (V + V_1 - V_2)}$$

$$= \frac{W}{(V_2 - V_1)}$$

Bulk density and pore space can also be calculated from the above data

$$\text{Bulk density} = \frac{\text{Weight of soil}}{\text{Volume of soil}} = \frac{W}{V}$$

$$\text{Pore space (per cent)} = \frac{(V + V_1 - V_2)}{V} \times 100$$

4. PORE SPACE

The per cent pore space is an important soil physical property and indicates the soil volume occupied by soil air or soil water. The total pore space is made up of micropores (capillary pores) and macropores (non-capillary pores). The micropores hold water while macropores are filled with water. In an ideal soil, the total pore space should be equally distributed between the two types of pores.

The total pore space of a medium textured soil is around 50%.

Sandy soils have lower porosity while clay and organic soils have higher porosities. Subsoils have lower pore space than the surface soils due to compaction. In a coarse textured soil, the proportion of macropores is greater while in a fine textured soil, the proportion of micropores is greater.

The pore space is calculated from the bulk density and particle density as indicated below:

Particle density of soil = D_p gm/ml

Bulk density of soil = D_b gm/ml

$$\text{Per cent pore space} = \left(1 - \frac{D_b}{D_p}\right) \times 100$$

or as indicated in the calculations of the experiment on particle density.

5. FIELD CAPACITY OF SOIL

Principle

Field capacity of a soil gives an estimate of the amount of water retained in soil after the rainfall and provides a rough estimate of the potential capacity of soil to supply water to plants. It is the percentage of water remaining in the soil after 2 or 3 days of free drainage of a saturated soil in the absence of evaporation and gives the amount of water in the micropores. This moisture is supposed to be retained in the soil at a tension of 0.33 bar (1/3 atmosphere) and will vary with the texture of soil and pore space distribution. In sandy soils where there are more macropores, water infiltrates rapidly and is lost as gravitational water and hence the proportion of water at field capacity is less. In soils with more of micropores (clay soils), the amount of water at field capacity will be more.

In this method, the soil is saturated with water and then allowed to drain for 48 hours without allowing any evaporation to take place. From the amount of water required to saturate the known weight of soil, the field capacity is estimated. This can be done with an undisturbed soil (*in situ*) and also with disturbed soils.

i. In situ analysis

Procedure

Select a spot 2m x 2m in area in the field. The water table should not be within two meters from the surface or from the layer whose field capacity is being determined. Bund the spot on all sides and add water to saturate the soil down to the depth whose field capacity is needed. Then cover the plot with a polythene sheet or some other material to prevent evaporation. After 48 hours, collect the soil samples with a tube augur and transfer it to previously weighed moisture box and find out the weight. Place the moisture boxes in an oven at 105°C for 24 hours, and then cool it in a desiccator and find out its weight.

Calculations

Weight of empty moisture box = W_1 gm

Weight of moisture box + moist soil = W_2 gm

Weight of moisture box with oven dry soil = W_3 gm

$$\text{Percentage of moisture at field capacity} = \frac{[W_2 - W_3]}{[W_3 - W_1]} \times 100$$

ii. Disturbed soil

Procedure

Select a 4 kg circular can with drainage holes at the bottom and place a layer of gravel of 2.5 cm thickness at the bottom. Fill up the can with the soil in small increments, tapping the can after each addition. Then add water to the container in small amounts till the soil is saturated as indicated by water seeping out through the drainage hole. Record the time taken for the water to infiltrate into the soil and also the quantity of water added. Add more water to the surface and collect drainage water at an interval of 1 hour. Continue it for 2 hours. Record the amount of water drained during the interval of one hour.

Cover the can with polythene sheet to prevent evaporation and leave it for 2 days. Then remove the cover and determine the moisture content after drawing a sample with the augur and keeping it in an air oven at 105°C for 24 hours. Also find out the depth of soil above the gravel surface and also the diameter of the can.

Calculations are the same as in the " in situ studies".

(Moisture content by weight ϕ_w can be converted into moisture content by volume ϕ_v using the formula

Per cent ϕ_v = per cent ϕ_w x D_b where D_b = bulk density of soil

For finding the moisture in terms of depth "ϕ depth" use the formula

ϕ depth = ϕ_v x d where ϕ_v is the volume of water per unit volume of soil, d = depth of soil).

iii. Pressure Plate Apparatus

Principle

The moisture retained by a saturated soil when subjected to 1/3 atmosphere pressure gives the field capacity and this can be estimated using a pressure plate apparatus with a ceramic plate and air pressure of 1/3 atmosphere.

Special equipments

Pressure plate apparatus, rubber soil retaining rings, moisture box.

Procedure

Moisten the ceramic plate (1/3 atmosphere) and fit it up in the pressure plate apparatus and trim the edges. Place the soil retaining ring on the membrane and put 25 gm of soil sample into the ring, level the sample in the ring and allow it to stand for atleast 16 hours with excess water on the plate. Remove the excess water, close the apparatus and admit air at pressure of 1/3 atmosphere.

After few hours, a marked decrease in rate of water flow will be observed. When the rate of flow decreases, apply a pressure differential of 4 lb/square inch to the rubber diaphragm at the top of soil chamber. (This diaphragm action holds the sample firmly in contact with the membrane and considerably hastens moisture extraction for fine textured soils that shrink appreciably. This is unnecessary for medium and coarse textured soil). Remove the sample after the equilibrium has been attained. Transfer the samples to moisture boxes and determine the moisture content by drying at 105°C.

Calculations

Weight of empty moisture box = W_1 gm

Weight of moisture box with moist soil = W_2 gm

Weight of moisture box with dry soil = W_3 gm

Weight of oven dry soil = $(W_3 - W_1)$

Weight of moisture = $(W_2 - W_3)$

$$\text{Percentage of moisture} = \frac{[W_2 - W_3]}{[W_3 - W_1]} \times 100$$

6. SATURATION CAPACITY

Principle

The saturation capacity of a soil gives the amount of water held by a soil when its macro- and micropores are completely saturated with water. It gives an approximate idea of total porosity. Saturation percentage is approximately four times the wilting percentage and two times the field capacity. Saturation percentage is determined with a Keen's box using disturbed soil. Keen's box is a cylindrical metal box 5.6 cm internal diameter and 1.6 cm height. The bottom of the box is perforated with holes of 0.75 mm diameter at 4 mm apart. Each box is made up of split brass wire ring which holds a filter paper in position over the perforated bottom.

Apparatus

Keen's box.

Procedure

Fit up a filter paper of appropriate size on the bottom screen of the Keen's box and weigh it. Then transfer 5-6 gm of the disturbed soil (passing through 2 mm) in small amounts, tapping the box gently every time to get a uniform packing till the box is filled a little over the top edge. Slick off the excess soil with a spatula. Add a small amount of soil and tap it and remove the excess. Place the box in a tray containing water atleast almost 0.5 cm above the screen of the Keen's box. Allow it to saturate till a shining film of water is obtained at the soil surface. Approximately 12-16 hours of capillary wetting would be needed for a fine textured soil. After complete saturation, remove the box and allow it to drain the excess water, wipe the sides of the box with a filter paper and weigh it. Dry it in an air oven at 105°C, cool it in a desiccator and weigh it again.

To find out the amount of water held by the filter paper, weigh ten filter papers similar to that placed in the Keen's box, after saturating it with water. From these weights, calculate the average amount of water held by a single filter paper and deduct the same from the weight observed.

Calculations

Weight of Keen's box and filter paper = W_1 gm

Weight of Keen's box + filter paper + saturated soil = W_2 gm

Weight of Keen's box + filter paper + oven dry soil = W_3 gm

Average weight of water held by the filter paper = W_4 gm

Saturation capacity per cent = W_S

$$= \frac{(W_2 - W_3 - W_4)}{(W_3 - W_1)} \times 100$$

7. INFILTRATION RATE OF SOIL

Principle

Infiltration refers to the downward entry of water into the soil and the rate at which it occurs is called the infiltration rate or water intake rate. It is rapid in the beginning and decreases with time and attains almost a steady value called "infiltration capacity" or "basic infiltration rate".

The infiltration rate is usually measured under field conditions and involves ponding of water on the surface or sprinkling of water to stimulate the rain. There is no generally accepted method suitable for all situations. The cylinder method involving the ponding of water is the most preferred method. A guard ring is used to check the lateral movement of water.

Apparatus

Two metal rings : Inner ring of 30 cm diameter and 20 cm in height and a guard ring of 50 cm diameter and 20 cm in length. One edge of the rings should be sharp so as to facilitate easy penetration into the soil.

Dividing plate : Steel plate 1.25 cm thick and 60 cm in diameter with legs to keep the plate on the rings. Hammer, watch, spade

Procedure

Drive the inner and guard rings in the soil by hammering on the driving plate placed on the rings until the top 10 cm of the rings remain above the soil surface. Place a gunny bag piece on the soil surface inside the inner ring and pour water both inside and outside the ring so that the water is ponded to a depth of 6 cm. There should not be any puddling of a soil during the addition of water. Both the rings should be pushed to the same depth in the soil and also the water inside and outside the inner ring should be approximately at the same level.

Record the initial level and continue recording the receding water level after suitable intervals. Note the observations after 2, 5, 20, 25, 30, 60, 90, and 120 minutes after infiltration has commenced. Note the depth of water infiltration (mm) and the rate of infiltration.

Classification of soil on the basis of infiltration rate

Class	Infiltration rate			Unit
Very rapid	>25.4			cm/hour
Rapid	12.7	–	25.4	cm/hour
Moderately rapid	6.3	–	12.7	cm/hour
Moderate	2.0	–	6.3	cm/hour
Moderately slow	0.5	–	2.0	cm/hour
Slow	0.1	–	0.5	cm/hour
Very slow	<0.1			cm/hour

8. HYGROSCOPIC COEFFICIENT

Principle

Air dry soil when exposed to a nearly saturated atmosphere (98% relative humidity) absorbs moisture and the per cent of moisture gives the hygroscopic coefficient.

Reagent

Sulphuric acid (3.3% by weight)

Procedure

Place 3.3% sulphuric acid in the lower part of the desiccator and keep a thin layer of soil on a watch glass inside the desiccator. Close the desiccator tightly and create a vacuum inside using a suction pump. Close the valve and maintaining the vacuum for 48 hours by which time equilibrium between the soil and the acid would have been attained. Remove the watch glass with the soil and weigh it. Dry the soil in an air oven at 105°C for 8 hours, cool it and weigh again. Calculate the moisture present as the hygroscopic coefficient.

Calculations

Weight of watch glass = W_1 gm

Weight of watch glass + moist soil = W_2 gm

Watch glass + oven dry soil = W_3 gm

Weight of oven dry soil = $(W_3 - W_1)$ gm

Amount of moisture = $(W_3 - W_2)$ gm

$$\text{Per cent of moisture} = \frac{(W_3 - W_2)}{(W_3 - W_1)} \times 100$$

9. WILTING PER CENT MOISTURE

Principle

The moisture retained by a saturated soil when exposed to 15 atmosphere tension gives the moisture per cent at wilting point.

Special apparatus

Pressure membrane apparatus, rubber soil retaining rings, moisture boxes.

Procedure

Moisten the cellulose membrane and fix up the same in the pressure membrane apparatus and trim the edges. Keep the soil retaining rings on the membrane and add 25 gm of soil into the ring and level it in the ring. Allow it to stand atleast for 16 hours with excess of water on the membrane and remove the excess water with a pipette. Close the pressure membrane apparatus and admit air into the soil chamber at 15 atmosphere pressure.

After few hours, the rate of water outflow from the soil decreases markedly. At this stage, a pressure differential of 4 lb/sq.inch may be applied to the rubber diaphragm at the top of the soil chamber. This diaphragm action holds the sample firmly in contact with the membrane and considerably hastens moisture extraction for fine textured soil that shrinks appreciably. This diaphragm is unnecessary for medium and coarse textured soils.

Remove the sample after 48 hours from the commencement of the extraction or when the readings on an outflow burette indicates that equilibrium has been reached. Then transfer the soil to the moisture boxes and determine the moisture content by drying it in an air oven at 105°C and calculate the moisture per cent at wilting point.

Calculations

Weight of moisture box = W_1 gm

Weight of moisture box + moist soil = W_2 gm

Weight of moisture box + dry soil = W_3 gm

Weight of oven dry soil = $(W_3 - W_1)$ gm

Amount of moisture = $(W_3 - W_2)$ gm

$$\text{Per cent moisture} = \frac{(W_3 - W_2)}{(W_3 - W_1)} \times 100$$

B. CHEMICAL PROPERTIES

10. Soil pH

The soil pH indicates the hydrogen or hydroxyl ion activity of the soil-water system and thereby indicates whether the soil is acidic, neutral or alkaline in reaction.

Number of techniques using pH strips, universal indicator, pH meters etc., have been adopted for finding out the soil pH.

i. By using pH strips

Principle

The pH strips are papers containing organic dyes (indicators) giving different colours at different hydrogen ion concentrations. They are of narrow or wide range and are differently coloured. These strips are dipped in soil suspensions and their resultant colours are matched with those on the colour chart provided on the box. The range of pH values and the changes in colour for different indicators is given below:

Indicators	Colour change	pH range
Thymol blue	Red - Yellow	1.2 - 2.8
Bromothymol blue	Yellow - Violet	2.8 - 4.6
Bromocresol green	Yellow - Blue	3.6 - 5.2
Methyl red	Red - Yellow	4.2 - 6.3
Bromocresol purple	Yellow - Purple	5.2 - 6.8
Bromothymol blue	Yellow - Blue	6.0 - 7.8
Cresol red	Yellow - Red	7.2 - 8.8

Barium sulphate may be added to settle the colloidal particles in the soil suspensions and the amount to be added depends upon the amount of colloidal particles present.

Reagent

pH strips of narrow range.

Procedure

Place a layer (1 cm thick) of barium sulphate at the bottom of the test tube, add soil to 3 cm thickness and fill up with distilled water up to 15 cm height. Shake thoroughly for 30 seconds and allow it to settle. Dip the pH strips in the soil suspensions and compare the colour of the strip with the colour given on the cover of each box of pH strips. In case it does not match, take the one that is close to the colour of the paper as the pH of the soil.

ii. BDH indicator

To a 20 gm soil sample kept in a conical flask, add 100 ml of water and 1 gm of barium sulphate. Shake the mixture for 10 minutes and allow it to stand for 30 minutes. Transfer 10 ml of the clear supernatent liquid into a test tube and add 2 to 3 drops of BDH universal indicator. Compare the colour developed with that of the standard and fix up the approximate pH of the extract.

iii. pH meter

Principle

A glass electrode in contact with the hydrogen ions of a solution acquires an electrical potential depending upon the concentration of hydrogen ions. This can be measured potentiometrically against some reference electrode which is usually a calomel electrode. The potential difference between the glass electrode and calomel electrode is expressed in pH units.

The pH meter measures the voltage developed by the combination of glass electrodes and reference electrodes and its scale is graduated in pH units instead of volts. The pH of the soil is measured with different soil to water ratio in different laboratories.

Equipments

pH meter

Procedure

To a weighed quantity of soil (20 gm) kept in a 150 ml beaker, add water equal to twice the weight of soil. This gives 1:2 ratio of soil:water (In certain laboratories 1:2.5, 1:5 ratios of soil: water are also adopted in which case the amount of water to be added should be modified suitably). Stir the contents with a glass rod and allow it to stand for 30 minutes. Again stir it and read the pH immediately using the pH meter. Care should be taken to immerse the electrodes properly in the suspensions and that too immediately and take the readings as otherwise the soil particles will settle and the sediment and supernatent liquid will have different pH value.

In case of soil water suspensions, the pH of acid, normal and alkaline soils range from less than 6.5, 6.5 to 8,7 and greater than 8.7 respectively.

iv. Soil pH in 0.01 M calcium chloride

Some laboratories recommend the use of 0.01 M calcium chloride solution instead of water for preparing the soil-water suspensions. The procedure is the same as described in (iii) but the pH value will be slightly less than that observed with water. Normally the pH of acid soils, calcareous soils and limed soils by this method will range from 4 to 5, 6.9 to 7.2 and around 6 respectively.

v. Soil pH using saturated soil paste

Some laboratories require the pH to be estimated using the saturated soil paste as this condition comes closer to that existing in a field near the plant roots.

Procedure

Add water in small amounts to a soil and mix it with a spatula. At the saturation point, the soil paste should glisten, flow slightly when the container is tipped or slide freely and cleanly of the spatula. After one hour, observe the soil paste. Free water should not collect on the soil surface and the paste should not stiffen markedly or loose its glistening appearance on standing.

After setting up the pH meter, insert the electrodes carefully in the soil paste and measure the pH. The electrodes may be moved a little to ensure the removal of water film around it and then measure the pH. Repeat it 3 to 4 times and take the average.

Precautions

i. Electrodes should not be rubbed strongly with hand or any materials as there are too fragile ones.

ii. Electrodes should not be allowed to dry. They should be kept in distilled water when not in use.

iii. Electrodes should not be allowed to remain in the test solution for a long time. After testing, it should be washed with distilled water immediately.

The pH meter commonly used is a glass electrode pH meter with calomel-reference electrodes including the salt bridge. Some latest models have single combined electrode and also have a digital reading unit to provide a direct reading instrument. As the pH meter is a potentiometer in principle, the pH scale should be calibrated before use with buffer solution of known pH values.

11. ELECTRICAL CONDUCTIVITY OF SOIL (EC of soils)

Principle

Ions are carriers of electricity and as the amount of soluble salts in a solution increases, the electrical conductivity (EC) of the soil - water system also increases giving rise to more of ion pairs on dissociation as in the case of dilute solutions. EC of soil is directly related to the soluble salt concentration of the soil and depends upon the ion activity and the conductance of the ions.

Electrical conductivity is measured with an AC salt bridge or electrical resistance bridge and conductivity cell having electrodes coated with platinum black. Instruments already calibrated are available for direct reading of conductivity of the solution in millimhos/cm at 25°C. The instrument is commonly known as SOLU BRIDGE using alternate current. The balance point is indicated by the MAGIC EYE. The conductivity cell consists of two platinum electrodes in the form of rectangular pieces fused on glass on one side and covered with black spongy platinum on the other. Two leads of wire connect it to the proper terminals of the solu bridge. The cell constant is usually marked on the cell. Though electrical conductivity is expressed in mhos/cm generally, as it is very low in soils and is in the order of mhos/cm x 10^{-3}, it is expressed as millimhos/cm. The approximate relationship between total soluble salts and electrical conductivity in millimhos/cm is:

$$\text{Per cent total soluble salts} = \frac{0.06 \times EC \times 10^3 \times SP}{100}$$

where SP = saturation percentage.

Reagent

Standard potassium chloride

Equipments

Solu bridge

Procedure

Shake 20 gm of soil with 40 ml of distilled water in 150 ml conical flask for one hour and allow it to stand (The clear extract from pH determination may also be used). Then take the supernatent liquid and determine the electrical conductivity using the conductivity bridge.

For this switch the conductivity bridge and wait till the magic eye is seen (illuminated). Wash the conductivity cell with distilled water, immerse it into the soil suspension in the case of dip type or suck the clear supernatent solution into the conductivity cell if it is a pipette type.

Rotate the meter knob until the black showed area in the magic eye is at the maximum or the widest. Take the reading for the conductivity in the scale meter. Multiply this reading by the cell constant noted on the top rim of the conductivity cell to arrive at the specific conductivity.

Calculations

Cell constant of the conductivity cell = X

Reading of the solubridge = Y mmhos/cm

Then electrical conductivity (specific conductivity) = X x Y mmhos/cm

Ratings

<0.8 mmhos/cm	Normal for all crops
0.9 - 1.6 mmhos/cm	Critical for salt sensitive crops
1.6 - 2.5 mmhos/cm	Critical for salt tolerant crops
>2.5 mmhos/cm	Injurious to all crops

12. TOTAL SOLUBLE SALTS (Gravimetric method)

Principle

The soluble salts in the soil is extracted by shaking a known quantity of soil with water and filtering the same. The soluble salts extracted is estimated by evaporating the same over a water bath to dryness.

Equipments

End over end shaker

Procedure

Transfer 100 gm of soil into 750 ml shaking bottle and add 200 ml of distilled water and shake it in an end over end shaker for one hour. This can be done manually also. Allow the suspension to settle for sometime and filter the same. Reject the first few ml of the extract if it is turbid and collect the rest for determination of soluble salts. Transfer 50 ml of the extract to previously weighed porcelain dish and evaporate it on a water bath. When the volume is around 2 ml, add 2 ml of 20% hydrogen peroxide and evaporate it on a water bath. After complete evaporation, dry it in an oven at 105°C for one hour, cool it in a desiccator and weigh.

Calculations

Weight of empty dish = W_1 gm

Volume of extract prepared = 200 ml

Volume of extract taken = V ml

Weight of dish + residue after drying = W_2 gm

$$\text{Per cent of total soluble salts} = (W_2 - W_1) \times \frac{200}{V}$$

13. INDIVIDUAL IONS IN SOIL WATER EXTRACTS

The soluble cations and anions commonly present in the soil extracts are Ca^{2+}, Mg^{2+}, Na^+, K^+, CO_3^{2-}, HCO_3^-, SO_4^{2-}, and Cl^-.

These ions are estimated by the same procedure described in the water analysis taking into consideration the amount of soil taken and extract collected and used.

14. ORGANIC CARBON CONTENT (Walkley and Black, 1934)

Principle

The soil organic matter is estimated by the Walkley and Black rapid titration method (1934). In this method the soil organic matter is extracted by digesting the soil with chromic acid and sulphuric acid making use of the heat of dilution of sulphuric acid. The excess chromic acid which is not reduced by the organic matter of the soil is determined with standard ferrous ammonium sulphate solution in the presence of sodium fluoride, phosphoric acid and diphenylamine solution indicator. At the end point, the colour of the suspension changes from violet through blue to bright green. Addition of sodium fluoride and phosphoric acid makes the colour change distinct (sharp) due to their flocculating property.

i. Oxidation of carbon

$2K_2 Cr_2O_7 + 8H_2SO_4$ -----> $2K_2 SO_4 + 2Cr_2 (SO_4)_2 + 8H_20 + 6(O)$

$3C + 6(O)$ -----> $3CO_2$

$2K_2 Cr_2O_7 + 8H_2SO_4 + 3C$ ----> $2K_2SO_4 + 2Cr_2 (SO_4)_3 + 8H_20 + 3CO_2$

ii. Titration with ferrous ammonium sulphate solution

$2\ FeSO_4 (NH_4)_2\ SO_4 . 6H_20$ ---- $2\ FeSO_4 + 2\ (NH_4)_2\ SO_4 + 12\ H_20$

$2\ FeSO_4 + H_2SO_4 + (O)$ ---- $Fe_2\ (SO_4)_3 + H_2O$

$2\ FeSO_4 (NH_4)_2\ SO_4 . 6H_20 + H_2SO_4 + (O)$ ------>

------> $2\ (NH_4)_2\ SO_4 + Fe_2\ (SO_4)_3 + 13\ H_2O$

iii. Action of diphenylamine indicator

$$2\ C_6 H_5 NH_5 C_6 H_5 \xrightarrow[-H_2O]{+O} 2\ (C_6 H_5 NHC_6 H_4) \xrightarrow[-H_2O]{+O} C_6 H_5 N - C_6 H_4 -$$

$C_6 H_4$ N diphenylbenzidine (violet) --> $C_6 H_3$ colourless

As organic matter on an average contains 58% organic carbon, the organic matter content is obtained by multiplying the organic carbon content by 1.724. This factor (100/58 = 1.724) is called Bemmlen factor.

Reagents

Ferrous ammonium sulphate (Mohr's salt) (0.5 N), diphenylamine indicator, concentrated sulphuric acid, sodium fluoride, orthophosphoric acid (85%), potassium dichromate (1.0 N)

Procedure

Weigh accurately 1 gm of soil ground to pass through 0.2 mm sieve and transfer it to a 500 ml conical flask. Add 10 ml of 1 N potassium dichromate and shake it for a while and then add 20 ml of concentrated sulphuric acid (containing 1.25% silver sulphate) and shake it two or three times. Allow it to stand for 30 minutes and then add 200 ml of distilled water, followed by 10 ml phosphoric acid or 0.5 gm of sodium fluoride and 1 ml of diphenylamine indicator. Titrate the contents with ferrous ammonium sulphate till the colour flashes from blue violet to green. Combination of phosphoric acid and sodium fluoride gives a sharper end point (If more than 7 ml dichromate solution is consumed, the analysis may be repeated with a lesser quantity of soil). Simultaneously run a blank without the soil.

Calculations

Weight of soil taken = W gm

Volume of 0.5 N ferrous ammonium sulphate solution used in blank = V_1 ml

Volume of 0.5 N ferrous ammonium sulphate solution used in titrating the excess nascent hydrogen = V_2 ml

Volume of 1.0 N potassium dichromate used for the titration

$$= \frac{V_1 - V_2}{2}$$

1 ml of 1N potassium dichromate = 0.003 gm of organic carbon

$$\frac{V_1 - V_2}{2} \text{ ml of 1 N potassium dichromate} = \left(\frac{V_1 - V_2}{2}\right) \times 0.003$$

Per cent of organic carbon in soil

$$= \frac{(V_1 - V_2)}{2} \times \frac{100}{W} \times 0.003 = X \text{ gm}$$

Per cent of organic matter = X x 1.724

Classification

Low = <0.5%; Medium 0.5 - 0.75%; High >0.75%.

ii. COLORIMETRIC METHOD (Datta et al., 1962)

Principle

The soil organic matter is oxidised by the dichromate-sulphuric acid and the intensity of green colour of chromium sulphate formed is measured in a colorimeter and from that the organic carbon content calculated.

Reagents

Potassium dichromate (1.0 N), concentrated sulphuric acid with 1.25 gm silver sulphate in 100 ml sucrose (anhydrous).

Procedure

Transfer one gram of soil passing through 0.2 mm sieve into a dry 100 ml conical flask and add 10 ml of 1 N potassium dichromate and mix it further. After keeping it for 30 minutes on an asbestos pad, centrifuge the contents and take the clear supernatent liquid and read the intensity of green chromium sulphate colour in a colorimeter using a red filter (660 mμ) after adjusting with the blank solution.

Standard calibration curve

Take one to 2.5 mg of anhydrous sucrose (AR) in a separate 100 ml dry conical flask, add 10 ml of 1 N potassium dichromate solution and 20 ml of concentrated sulphuric acid in that order. Shake the flask for a while and keep it for 30 minutes. Then read the colour of the solution in a colorimeter at 660 mμ after adjusting the blank to zero. Draw a graph with concentration of carbon (in sucrose) and the colorimeter readings and calculate the factor for sucrose. In a Klett-Summerson colorimeter, the factor comes to 0.0042 with slight variation.

(Instead of centrifuging, after the addition of potassium dichromate and sulphuric acid to the soil, the flask can be kept overnight and the supernatent solution carefully decanted off for colorimetric analysis. This reduces the possibility of corrosion to the centrifuge due to the spilling of concentrated acid and also will enable handling a large number of samples. If chromic acid crystals appear in cold climate, addition of 10 ml of water to each flask after 30 minutes reaction can be adopted. In that case the standard curve also should be prepared in a similar manner).

Calculations

Organic carbon content = Colorimetric reading (R) x factor from the standard curve.

15. LOSS ON IGNITION

Principle

When a known weight of soil is ignited at around 600°C, the loss in weight is due to the destruction of organic matter, calcium and magnesium carbonate and loss of moisture. The sum total of all these are considered as loss on ignition.

Apparatus

Silica basin or crucible

Procedure

Heat an empty silica basin (or crucible) for about 30 minutes on a rose head flame, cool it in a desiccator and find out its weight. To this add about 20 gm of soil and find out its weight accurately. Ignite the contents over a rose head flame for about 8 hours by which time all the organic matter and combined water would have been removed. Transfer the basin with the soil to a desiccator, allow it to cool and record the weight. The loss in weight represents the loss on ignition together with the moisture loss at 105°C.

Calculations

Weight of empty silica basin = W_1 gm

Weight of silica basin + soil = W_2 gm

Weight of soil = $W_2 - W_1$ gm

Weight of silica basin + soil (after ignition) = W_3 gm

Loss on ignition = $W_2 - W_3$ gm

$$\text{Percentage of loss on ignition} = \frac{(W_2 - W_3)}{(W_2 - W_1)} \times 100$$

Loss on ignition on moisture free basis =

$$= \left[\frac{(W_2 - W_3)}{(W_2 - W_1)} \times 100\right] - \left\{M \times \left[\frac{100}{(100 - M)}\right]\right\}$$

(M = Moisture per cent)

16. CHEMICAL ANALYSIS OF SOIL USING HYDROCHLORIC ACID EXTRACT

a. Hydrochloric acid extract

Principle

In this method, a known weight of soil is ignited to destroy the organic matter and the ignited soil is digested with hydrochloric acid to bring the mineral constituents into solution. The extract thus obtained is used for further analysis.

Reagents

Hydrochloric acid (1:1).

Procedure

Transfer an accurately weighed quantity of soil (20 gm) into a silica basin and ignite it over a rose head flame for 6 hours till all the organic matter is destroyed. Then transfer the ignited soil to a 250 ml conical flask using minimum quantity of water. Add 60 ml of 1:1 hydrochloric acid, cover the mouth of the conical flask with a funnel and digest it over a sand bath for 8 hours. At the end of the period, dilute it to 200 ml and allow it to stand overnight.

To prevent loss of material through the sides of the conical flask while filtering, smear the mouth of the conical flask (outside) with a thin layer of Vaseline. Then filter the solution through Whatman No. 3 filter paper into a 500 ml volumetric flask. Transfer all the materials to the filter paper and wash it with hot water till the filtrate is free of chloride. Cool the extract and then make up the volume with cold distilled water, shake thoroughly and keep it for further analysis.

b. Acid Insolubles

The material remaining on the filter paper after preparation of the hydrochloric acid extract is called the acid insolubles. The amount of the acid insolubles present in the soil is estimated by transferring the dry residue to a previously weighed silica basin (or silica crucible), igniting the same over a rose head flame to constant weight, cooling the basin (or crucible) in a desiccator to the room temperature and weighing it. From the weight thus obtained the per cent of acid insolubles in the soil was calculated.

Calculations

Weight of empty silica basin A = W_1 gm

Weight of silica basin A + soil before preparation of hydrochloric acid extract = W_2 gm

Weight of soil taken = $W_2 - W_1$ gm

Weight of empty silica basin B = W_3 gm

Weight of silica basin B + acid insolubles= W_4 gm

Weight of acid insolubles = $(W_4 - W_3)$ gm

$$\text{Per cent of acid insolubles} = \frac{(W_4 - W_3)}{(W_2 - W_1)} \times 100$$

Per cent of acid insolubles on moisture free basis

$$= \frac{(W_4 - W_3)}{(W_2 - W_1)} \times 100 \times \frac{100}{(100 - M)}$$

c. IRON AND ALUMINIUM OXIDES

Principle

The iron and aluminum present in the hydrochloric acid extract are estimated by precipitating them as their hydroxides by the addition of ammonium hydroxide. The hydroxides are then converted into the oxide form by igniting the same. From the weight of the dried precipitate, the amount of iron and aluminium oxides present in the hydrochloric acid extract is estimated.

Reagents

Ammonium chloride (solid), dilute ammonium hydroxide.

Procedure

Pipette out 50 ml of hydrochloric acid extract into a 250 ml beaker and add 1-2 gm of solid ammonium chloride to make the precipitate granular. Cover the beaker with watch glass and boil. Remove the beaker and add dilute ammonium hydroxide to the hot solution carefully till the solution is alkaline to litmus. Boil again for 4 to 5 minutes and filter it through Whatman No. 3 filter paper, collecting the filtrate in a 500 ml beaker. Transfer all the precipitate to the filter paper using hot water and also wash it with hot water. Retransfer the precipitate to the beaker with a jet of warm water, boil and filter again. Repeat this number of times (5-7 times) till the filtrate is free of chloride (Preserve the filtrate for estimation of calcium and potassium).

Dry the precipitate with the filter paper in an electric oven, transfer it to a weighed silica crucible and ignite it to constant weight. From the weight of the precipitate, calculate the percentage of sesquioxides.

Calculations

Weight of soil taken = W gm

Volume of hydrochloric acid extract prepared = 500 ml

Volume of hydrochloric acid extract taken for analysis = 50 ml

Weight of silica crucible alone = W_1 gm

Weight of silica crucible + ignited precipitate (sesquioxides) = W_2 gm

Weight of sesquioxides = $(W_2 - W_1)$ gm

Per cent of sesquioxides on moisture free basis

$$= (W_2 - W_1) \times \frac{500}{50} \times \frac{100}{W} \times \frac{100}{(100 - M)}$$

d. IRON ALONE

Principle

The iron in the hydrochloric acid extract is converted into ferric form by the addition of concentrated sulphuric acid. The ferric iron is reduced to the ferrous form by adding dilute sulphuric acid and zinc granules. The ferrous iron is reoxidised to ferric form by titration with standard potassium permanganate. From this the content of iron is estimated.

Hydrochloric acid extract

$$Fe + 6H_2SO_4 \text{ -----> } Fe_2(SO_4)_2 + 6H_20 + 3SO_2$$

$$Zn + H_2SO_4 \text{ -----> } ZnSO_4 + 2(H^+)$$

$$Fe_2(SO_4)_3 + 2(H^+) \text{ ------> } 2FeSO_4 + H_2SO_4$$

Titration

a. $2KMnO_4 + 3H_2SO_4 \text{ -----> } K_2SO_4 + 2MnSO_4 + 3H_2O$

b. $10FeSO_4 + 5H_2SO_4 \text{ ----> } 5Fe_2(SO_4)_3$

Adding together the last two equations

$$2KMnO_4 + 8H_2SO_4 + 10FeSO_4$$

$$\text{-------> } K_2SO_4 + 2MnSO_4 + 5Fe_2(SO_4)_3 + 8H_20$$

From this equation, it could be seen that $2KMnO_4 = 5(O) = 10FeSO_4$ or two molecules of ferrous sulphate are oxidised by one atom of oxygen. So the equivalent weight of ferrous sulphate is its molecular weight.

Reagents

Concentrated sulphuric acid, zinc granules, sulphuric acid (1:4), sodium carbonate (solid), potassium or ammonium thiocyanate (2%) solution, potassium permanganate (0.1 N), glass wool.

Procedure

Pipette 25 ml of hydrochloric acid extract into a porcelain basin and evaporate it to dryness over a water-bath. When it is completely dry, add 1-2 ml of concentrated sulphuric acid and evaporate almost to dryness. The residue should be white showing all iron has been converted into ferric sulphate. Transfer the residue to 250 ml conical flask using hot water and add 50 ml of 1:4 sulphuric acid and zinc granules. Cover the mouth of conical flask with a funnel. Warm if necessary, to start the reaction and allow it to stand for 30 minutes for completion of reduction. Test for the completion of reduction using ammonium thiocyanate drops kept on a porcelain tile against a drop of the solution. The indicator should not turn blood red colour.

Filter this through glass wool into a 250 ml conical flask containing a pinch of sodium carbonate. Wash the original flask with hot water and collect the washings in 250 ml conical flask. Titrate it immediately against 0.1 N potassium permanganate till the appearance of permanent pink colour.

Calculations

Weight of soil taken = W gm

Volume of hydrochloric acid extract prepared = 500 ml

Volume of hydrochloric acid taken = 25 ml

Volume of 0.1 N potassium permanganate used = V ml

1 ml of 0.1 N potassium permanganate = 0.008 gm Fe_2O_3

Amount of iron present in the aliquot taken (Fe_2O_3)

$$= 0.008 \times V \text{ gm}$$

Amount of Fe_2O_3 present in 500 ml of extract

$$= W \text{ gm} = 0.008 \times V \times \frac{500}{25}$$

This is present in W gm of soil

$$\text{Per cent of } Fe_2O_3 \text{ in the soil } = 0.008 \times V \times \frac{500}{25} \times \frac{100}{W}$$

Per cent of Fe_2O_3 in the soil on moisture free basis

$$= 0.008 \times V \times \frac{500}{25} \times \frac{100}{W} \times \frac{100}{(100 - M)}$$

e. ALUMINIUM (Al_2O_3)

The aluminium content (Al_2O_3) is obtained by subtracting the iron content from the total iron and aluminium present in the soils.

Per cent of Al_2O_3 = (Total $Fe_2O_3 + Al_2O_3$) - (Fe_2O_3)

f. CALCIUM

Principle

Calcium is precipitated as calcium oxalate in acetic acid medium by the addition of ammonium oxalate and the precipitate is washed free of chloride, dissolved in dilute sulphuric acid and titrated against potassium permanganate.

Reagents

Ammonium chloride (solid), dilute ammonium hydroxide, dilute sulphuric acid, dilute acetic acid, saturated ammonium oxalate, potassium permanganate (0.1 N)

Procedure

Concentrate the filtrate from iron and aluminium oxide estimation and make up to a known volume (250 ml). Pipette out 50 ml of the filtrate into a 400 ml beaker, add 2 gm of ammonium chloride (solid) and a piece of red litmus. Add dilute ammonia till it is alkaline followed by dilute acetic acid till it is red again. Heat the contents and proceed as in the procedure for calcium estimation in plant analysis.

Calculations

Weight of soil = W gm

Volume of hydrochloric acid extract prepared = 500 ml

Volume of hydrochloric acid extract taken for sesquioxide estimation = 50 ml

Volume of filtrate made up = 250 ml

Volume of filtrate taken for calcium estimation = 50 ml

Volume of 0.1 N permanganate used = V ml

1 ml of 0.1 N permanganate = 0.002 gm Ca

V ml of 0.1 N permanganate (this is present in 50 ml)
$= 0.002 \times V$ gm

Hence in 250 of the filtrate (= 50 ml of hydrochloric acid extract) $= 0.002 \times V \times \frac{250}{50}$

500 ml of hydrochloric acid contains (= W gm of soil)

$$= 0.002 \times V \times \frac{250}{50} \times \frac{500}{50}$$

Per cent calcium (Ca) $= 0.002 \times V \times \frac{250}{50} \times \frac{500}{50} \times \frac{100}{W}$

Moisture per cent = M

Per cent of calcium on moisture free basis

$$= 0.002 \times V \times \frac{250}{50} \times \frac{500}{50} \times \frac{100}{W} \times \frac{100}{100 - M}$$

g. MAGNESIUM

Principle

Magnesium is precipitated as magnesium ammonium phosphate by the addition of sodium phosphate in an alkaline medium. The precipitate is filtered, washed free of chlorides, ignited and weighed as magnesium pyrophosphate.

$MgCl_2 + Na_2HPO_4 + NH_4OH \text{ -----> } Mg(NH_4)PO_4 + 2\,NaCl + H_2O$

$2\,MgNH_4PO_4 \text{ ----- } Mg_2P_2O_7 + 2\,NH_3 + H_2\,0$

Reagents

Ammonium chloride (solid), dilute ammonium hydroxide (1:4) and (1:7), disodium phosphate (10%).

Procedure

Pipette 50 ml of the filtrate from the sesquioxide estimation in a 400 ml beaker and add 1 gm of solid ammonium chloride followed by 1:4 ammonium hydroxide till the red litmus turns blue. Add slowly 10 ml of disodium phosphate (10%) solution drop by drop and stir the solution vigorously without touching the sides of the beaker. Allow it to stand for 30 minutes, preferably overnight to ensure complete precipitation. Filter through Whatman No. 42 filter paper by decantation and wash with dilute ammonia (1 : 7) till the filtrate runs free of chloride. Transfer the precipitate with the filter paper to a weighed silica basin and dry it in an electric oven. Ignite the dried precipitate over a rose head flame. If the ash is not white, allow the crucible to cool and add one or two drops of concentrated nitric acid. heat slowly first and strongly for another 30 minutes, cool in a desiccator and record the weight. Repeat the ignition till constant weight is obtained. From the weight of magnesium pyrophosphate obtained, calculate the percentage of magnesium in the soil.

Calculations

Weight of soil = W gm

Volume of hydrochloric acid extract prepared = 500 ml

Volume of hydrochloric acid taken for sesquioxide estimation = 50 ml

Volume of filtrate from the sesquioxide estimation made up to = 250 ml

Volume of filtrate taken for magnesium estimation = 50 ml

Weight of silica basin = W_1 gm

Weight of silica basin + magnesium precipitate ($Mg_2P_2O_7$) = W_2 gm

Weight of precipitate alone = $(W_2 - W_1)$ gm

1 gm of $Mg_2P_2O_7$ contains 2 Mg atoms

222 gm of $Mg_2P_2O_7$ = 2 x 12 = 24 gm Mg

$$(W_2 - W_1) \text{ gm of } Mg_2P_2O_7 = \frac{24}{222} \times (W_2 - W_1)$$

This is present in 50 ml of the sesquioxide filtrate

$$250 \text{ ml of the filtrate} = \frac{24}{222} \times (W_2 - W_1) \times \frac{250}{50}$$

This is present in 50 ml of hydrochloric acid extract

500 ml of hydrochloric acid extract (= W gm of soil)

$$= \frac{24}{222} \times (W_2 - W_1) \times \frac{250}{50} \times \frac{500}{50}$$

Amount of magnesium in 100 gm of soil

$$= \frac{24}{222} \times (W_2 - W_1) \times \frac{250}{50} \times \frac{500}{50} \times \frac{100}{W}$$

Per cent of magnesium on moisture basis

$$= \frac{24}{222} \times (W_2 - W_1) \times \frac{250}{50} \times \frac{500}{50} \times \frac{100}{W} \times \frac{100}{(100 - M)}$$

(M = moisture percentage)

h. POTASSIUM

Principle

The filtrate from the sesquioxide estimation is fed into a flame photometer and the intensity of emission is measured and from this the concentration of potassium in the soil is calculated.

Special equipment

Flame photometer

Procedure

Take 10 ml of the filtrate obtained from sesquioxide estimation and make it up to a known volume (250 ml) in a small vial and estimate the potassium content in a flame photometer as described in either water analysis or plant analysis.

Calculations

Weight of soil taken = W gm

Volume of hydrochloric acid extract prepared = 500 ml

Volume of hydrochloric acid taken for sesquioxide estimation = 50 ml

Volume of sesquioxide filtrate made up = 250 ml

Reading observed in the flame photometer = X

"ppm" of K equivalent to the reading (X) = Y

1 ml of the filtrate contains "Y" micrograms of K=$Y \times 10^{-6}$ gm K

250 ml of the filtrate (50 ml of hydrochloric acid extract)

$$= \frac{Y}{10^6} \times 250 \text{ gm K}$$

500 ml of hydrochloric acid extract (W gm of soil)

$$= \frac{Y}{10^6} \times 250 \times \frac{500}{50}$$

Per cent of potassium in soil = $\frac{Y}{10^6} \times 250 \times \frac{500}{50} \times \frac{100}{W}$

(Moisture per cent = M)

Per cent of potassium on moisture free basis

$$= \frac{Y}{10^6} \times 250 \times \frac{500}{50} \times \frac{100}{W} \times \frac{100}{(100 - M)}$$

i. PHOSPHORUS

Principle

Phosphorus in the hydrochloric acid extract is precipitated in nitric acid medium as ammonium phosphomolybdate and this is estimated by volumetric method.

Reagents

Hydrochloric acid (1:1), nitric acid (1:1), concentrated ammonium hydroxide, concentrated nitric acid, ammonium nitrate (solid), ammonium molybdate solution (20%), potassium hydroxide (0.1619 N), nitric acid (0.1619 N), phenolphthalein.

Procedure

As soluble silica in hydrochloric acid extract interferes with phosphorus estimation, this has to be converted into insoluble silica and eliminated. This is done by the following procedure:

Pipette out 200 ml of hydrochloric acid extract into a 400 ml beaker and evaporate to a small volume. Transfer it to a porcelain basin using hot water and evaporate to dryness on a water bath. Keep the basin in an air oven at 105 - 110°C for 3 hours to dehydrate the silica and render it insoluble (or this can be done by roasting the precipitate over a low flame till the residue becomes brown taking care to see that no residue spurts out). Dissolve the residue again in a small quantity of 1:1 hydrochloric acid and evaporate it to dryness over a water bath. Dissolve the residue again in 1:1 nitric acid, adding just sufficient amount of nitric acid to dissolve the same. Keep it overnight to allow the insoluble silica to settle down. Filter through Whatman No. 42 filter paper and wash the residue in the silica basin and on the filter paper with small quantities of 1:4 nitric acid till no yellow colour is left either in the basin or on the filter paper. Collect the filtrate and washings in 250 ml beaker taking care to see that the total volume does not exceed 100 ml.

Make the extract alkaline with concentrated ammonium hydroxide and then acidic with concentrated nitric acid and proceed to estimate the phosphorus as in the case of plant materials.

Calculations

Weight of soil taken = W gm

Volume of hydrochloric acid extract prepared = 500 ml

Volume of hydrochloric acid extract taken = 200 ml

Volume of 0.1619 N potassium hydroxide added = V_1 ml

Volume of 0.1619 N nitric acid used for back titration = V_2 ml

Volume of 0.1619 N potassium hydroxide used to dissolve the precipitate = $(V_1 - V_2)$ ml

1 ml of 0.1619 N potassium hydroxide = 0.0005 gm P_2O_5

$(V_1 - V_2)$ ml of 0.1619 N potassium hydroxide = $0.0005 \times (V_1 - V_2)$

This is present in 200 ml of hydrochloric acid extract

500 of hydrochloric acid extract contains (= W gm of soil)

$$= 0.0005 \times (V_1 - V_2) \times \frac{500}{200}$$

$$\text{Per cent of } P_2O_5 = 0.0005 \times (V_1 - V_2) \times \frac{500}{200} \times \frac{100}{W}$$

Per cent of P_2O_5 in the soil on moisture free basis

$$= 0.0005 \times (V_1 - V_2) \times \frac{500}{200} \times \frac{100}{W} \times \frac{100}{(100 - M)}$$

(Moisture per cent = M)

17. TOTAL NITROGEN

Principle

The nitrogen in the soil is converted into ammonium sulphate by digestion with sulphuric acid-salicylic acid mixture in the presence of sodium thiosulphate-potassium sulphate and copper sulphate. The ammonium sulphate that is formed, is digested with excess alkali and the liberated ammonia is collected in known excess of acid and estimated.

Reagents

Sulphuric acid - salicylic acid mixture, sodium thiosulphate, potassium sulphate - copper sulphate mixture (10:1), potassium hydroxide (0.1 N), sulphuric acid (0.1 N), sodium hydroxide (40%), sodium sulphide (10%), methyl red indicator.

Procedure

Weigh accurately a known quantity of soil (10 gm) and

proceed as in the estimation of total nitrogen in mixed fertilizer (or manures) or plant materials.

Calculations

Same as that given in the nitrogen estimation of manures or plant materials.

18. CATION EXCHANGE CAPACITY OF SOIL

Principle

The cation exchange capacity measures the total quantity of negative charges per unit weight of soil and is determined by leaching a known amount of soil by neutral salt solution (neutral normal ammonium acetate) and estimated by distilling the quantity of NH^+ ions - cations adsorbed by the soil. Cation exchange capacity gives a direct measure of the capacity of the soil to hold most of the plant nutrients against leaching losses of easily available forms of plant nutrients.

Reagents

Neutral normal ammonium acetate, alcohol (60%), ammonium chloride (solid), sulphuric acid (0.1 N), potassium hydroxide (0.1 N), methyl red indicator, red litmus.

Procedure

Transfer 10 gm of air dried soil into a 250 ml beaker and add 50 ml of neutral normal ammonium acetate solution, stir well and keep it overnight, covering it with a watch glass. Then filter it through Whatman No. 33 filter paper and leach it further with portions of 25 to 30 ml of ammonium acetate 6 - 8 times, collecting the filtrate in a beaker. Transfer the soil in the beaker to the filter paper using 60% alcohol and wash it with the same solvent (alcohol) until the filtrate is free of chloride. Remove the filter paper with the soil cautiously and transfer it to nitrogen distillation flask. Add distilled water till the flask is half full.

Keep an ice tumbler (beaker) with 25 ml of 0.1 N sulphuric acid and 1 - 2 drops of methyl red indicator at the delivery end of the distillation apparatus.

To the distillation flask add few drops of paraffin, few bits of porcelain and 10 ml of 40% sodium hydroxide and close the flask immediately and distill the ammonia. Collect the distillate till it is free of alkali (test with red litmus paper). Wash the end of the delivery tube with distilled water into the ice tumbler and titrate the excess acid using 0.1 N potassium hydroxide, till the appearance of light yellow colour from red colour.

Calculations

Weight of soil = W gm

Volume of 0.1 N sulphuric acid taken = V_1 ml

Volume of 0.1 N potassium hydroxide used = V_2 ml

1 ml of 0.1 N sulphuric acid = 0.1 meq

Cation exchange capacity

$$= (V_1 - V_2) \times 0.1 \times \frac{100}{W} \text{ meq/100 gm soil}$$

AVAILABLE NUTRIENTS

19. AVAILABLE NITROGEN

The available nitrogen content of soil is determined by number of methods involving the determination of mineral forms of nitrogen (NH_4-N and NO_3-N) produced on incubation of soil under aerobic and anaerobic conditions. Chemical methods for determination of soil available nitrogen involves either direct or indirect estimation of total nitrogen or measurement of a fraction of easily hydrolysable nitrogen by employing solution of dilute acid or alkali. Distillation with alkaline potassium permanganate has been used for estimating oxidisable and reactive forms of soil nitrogen.

i. ALKALINE PERMANGANATE METHOD (Subbiah and Asija, 1954)

Principle

This method involves distilling the soil with alkaline permanganate solution and determining the liberated ammonia. This method takes into account the reserve nitrogen and can be used for the determination of total nitrogen supplying capacity of a soil.

Alkaline potassium permanganate is a mild oxidising agent and can extract oxidisable fractions of organic nitrogen. In this method, a known weight of soil is treated with an excess of alkaline permanganate and the evolved ammonia is absorbed in standard acid and the excess amount of acid is estimated by back titrating with standard alkali with methyl red as indicator as in all other nitrogen estimations.

Reagents

Potassium permanganate (0.32%), sodium hydroxide (2,5%), methyl red indicator, sodium hydroxide (0.02 N), sulphuric acid (0.02 N), mixed indicator (methyl red-bromocresol green), boric acid (2%)

Procedure

Take 20 gm of soil in 800 ml dry Kjeldahl flask and to this add 20 ml of water, followed by 100 ml of each of 0.32% potassium permanganate and 2.5% sodium hydroxide solution. If there is no frothing, add 1 ml of liquid paraffin and then a few glass beads to prevent bumping.

At the distillation end, keep a conical flask containing 20 ml of boric acid with mixed indicator and then start distillation. Collect the distillate for 30 minutes. Then titrate it against 0.02 N sulphuric acid to the original colour (pink to green colour).

Or the distillate can be collected in excess of 0.02 N sulphuric acid (25 ml) with methyl red as indicator and then the excess acid titrated back with 0.02 N potassium hydroxide.

Calculations

Weight of soil = 20 gm

Volume of 0.02 N sulphuric acid taken = 25 ml

Volume of 0.02 N potassium hydroxide used = X ml

Volume of 0.02 N sulphuric acid used = (25 - X) ml

1 ml of 0.02 N sulphuric acid = 0.00028 gm N

$$\text{Per cent of available nitrogen} = 0.00028 \times (25 - X) \times \frac{100}{20} = Y$$

Available nitrogen in ppm = (N %) x 10,000

Available nitrogen in kg/ha = (N %) x 22,400=(25 - X) x 31.36

(Instead of 0.02 N sulphuric acid, the evolved ammonia can be collected in boric acid (20 ml) with mixed indicator as indicator and the absorbed ammonia estimated by titrating with 0.02 N sulphuric acid.

The available nitrogen in kg/ha = X x 31.36 kg/ha

where X = volume of 0.02 N sulphuric acid used)

Classification

Low	<272	kg/ha.
Medium	272- 544	kg/ha and
High	>544	kg/ha.

ii. NITROGEN MINERALISATION UNDER AEROBIC CONDITIONS (Ammonification and nitrification) (Keeney and Bremner, 1966)

Procedure

Weigh 10 gm of soil and mix it with acid washed quartz sand (30 - 60 mesh), transfer to a 250 ml wide mouth bottle containing 6 ml of distilled water and mix thoroughly. The bottle may be closed with a rubber stopper having a central hole for aeration. Incubate it at 30°C for 14 days and at the end of the period add 100 ml of 2 N potassium chloride and shake the suspension for one hour on a mechanical shaker and then filter it through Whatman No. 1 filter paper.

Transfer 20 ml of clear filtrate into the distillation flask of nitrogen distillation set, add 80 ml of water, 0.2 gm of magnesium oxide (freshly ignited) and Devarda's alloy and proceed as in the estimation of nitrate nitrogen of fertilizers.

iii. ANAEROBIC INCUBATION METHOD (Warring and Bremner, 1964: Subbiah and Bajaj, 1965)

Procedure

Place 12.5 ml of distilled water in 150 ml conical flask and add 5 gm of air dry soil, stopper the bottle and incubate it at 40°C for 7 days. At the end of the incubation, add 25 ml of 4 N potassium chloride or 50 ml of 2 N potassium chloride and shake it for 15 minutes, filter the suspension immediately and distill the filtrate with 0.5 gm magnesium oxide (freshly ignited) and collect the distillate in 20 ml 2% boric acid with mixed indicator. Titrate it against standard 0.02 N sulphuric acid and determine the ammonia evolved. From this, deduct the ammonia evolved by the soil without incubation.

20. AVAILABLE PHOSPHORUS

Number of chemical soil tests have been proposed for estimating available phosphorus in soil based on either the concept of stimulating the action of plant roots or to obtain a measure of readily soluble inorganic forms to represent the available phosphorus in the soil. In these methods, the nutrient is extracted with the extractant according to the specified soil, soil - water ratio and time of shaking. The extracted phosphorus is estimated colorimetrically.

i. OLSEN'S METHOD (Olsen et al., 1954)

Principle

A 0.5 M solution of sodium bicarbonate adjusted to pH 8.5 is designed to control the ionic activity of calcium through the solubility product of calcium carbonate in the case of neutral and calcareous soils. In this method the most reactive forms of phosphorus from the phosphate of iron, aluminium and calcium in the soil are extracted. The extractant raises the carbonate activity of the soil and decreases the calcium activity. The Darco G-60 added adsorbs the dispersed organic matter and makes the filtrate colourless. The soluble phosphate forms heteropoly complex with the molybdate ions giving a faint yellow colour. This on reduction with stannous chloride gives blue colour and the intensity of the blue colour which is proportional to the concentration, is measured in a colorimeter at 660 mμ using a red filter.

Reagents

Sodium bicarbonate (0.5 M) adjusted to pH 8.5, chloromolybdic reagent (1.5), Dickman and Bray's reagent having excess of acid, stannous chloride, Darco G-60 (or activated charcoal).

Special equipment

Colorimeter with a red filter

Procedure

Transfer 2.5 gm of soil into a 100 ml conical flask, add a little Darco G-60 (free of phosphorus) and 50 ml of Olsen's reagent (sodium bicarbonate 0.5 M adjusted to pH 8.5), shake it for 30 minutes in a shaker and filter the contents immediately through a dry Whatman No. 1 filter paper into a clean and dry beaker.

Estimate phosphorus in the extract colorimetrically by Dickman and Bray's (excess acid) method. Pipette 5 ml of the above soil extract into a 25 ml volumetric flask, add 5 ml of Dickman and Bray's reagent drop by drop with constant stirring till the effervescence ceases. Dilute it to 22 ml and add one ml of diluted stannous chloride solution and then make up to mark. Shake the solution and read the intensity of the blue colour in colorimeter using 660 mμ filter just after 10 minutes. As the colour fades after sometime, take the colorimetric readings quickly and also run a blank. The concentration of phosphorus in the test solution may be read from the standard curve.

Calculations

Weight of soil = 2.5 gm

Volume of Olsen's reagent added = 50 ml

Volume of extract taken for analysis = 5 ml

Final volume of solution = 25 ml

Total dilution = 10 x 5 = 50

Transmittance of the test solution = T

Concentration of phosphorus as observed in the standard curve = X ppm

Available P in ppm/ha = X x 50

Available P in kg/ha = ppm x 2.24

Classification

Low	< 12.4	kg/ha
Medium	12.4 - 22.4	kg/ha
High	> 22.4	kg/ha

b. Ascorbic acid method (Watanabe and Olsen, 1965)

Colour could be developed by using ascorbic acid also as indicated below: Pipette out 5 ml of the Olsen's extract into a 25 ml volumetric flask and carefully acidify the same to a pH 5 with 5 N sulphuric acid. This is done by taking 5 ml of 0.5 N sodium bicarbonate in a separate flask and determining the volume of acid to be added to bring it to pH 5 using p-nitrophenol indicator whose yellow colour disappears at that pH.

After adjusting the pH, dilute it to 20 ml with distilled water and add 4 ml of ascorbic acid solution. Make up to the mark and shake the contents well. After 10 minutes, read the intensity of blue colour in a colorimeter using 730-840 mμ filter. The colour is fairly stable. Run a blank. The calculations are the same as in the previous one.

c. Bray No. 1 method (for moderate to strong acid soils) (Bray and Kurtz, 1945).

Principle

This method is primarily meant for the soils which are moderately to strongly acidic in nature (pH). The Bray's No 1 reagent consists of 0.03 N ammonium fluoride in 0.025 N hydrochloric acid.

Reagents

Bray's No. 1 reagent, boric acid (5%), Dickman and Bray's reagent, stannous chloride

Special equipment

Colorimeter with red filter (660 mμ)

Procedure

Transfer an accurately weighed quantity of soil (5 gm) to a 100 ml conical flask, add 50 ml of Bray's No. 1 reagent and shake it for 5 minutes and filter. To avoid interference of fluoride, add 7.5 ml of boric acid (5%) to 5 ml of extract if necessary. Estimate the colour colorimetrically after developing the colour by the addition of Dickman and Bray reagent and stannous chloride. Read the colour intensity at 660 mμ (red filter) just after 10 minutes. From the standard curve, read the concentration of phosphorus in the extract.

Calculations

Weight of soil taken = W gm

Volume of Bray's reagent added = 50 ml

Volume of extract taken for analysis = 5 ml

Volume of solution made up = 25 ml

Transmittance of test solution = T

Concentration of P as observed from the standard curve
= X ppm

Available P per hectare in ppm = X

Available P per hectare in kg/ha = X x 4.48 =

$$\frac{[\mu g \text{ P in the aliquot (X)} \times \text{total volume of the extractant} \times 2.24 \times 10^6]}{[\text{Volume of aliquot} \times \text{weight of soil} \times 10^6]}$$

21. AVAILABLE POTASSIUM

Principle

Though all forms of potassium are in equilibrium, still the available potassium in soil generally represents the sum of water soluble and exchangeable potassium. The readily exchangeable and water soluble potassium is extracted with neutral ammonium acetate (CH_3 $COONH_3$) solution. The ammonium ion in the ammonium acetate exchanges with exchangeable potassium ions of the soil. As the NH_4 - ions hold the highly charged layers together just as potassium does, the release of non-exchangeable potassium is prevented during the extraction and this provides a clear separation. The extracted potassium is estimated with the help of flame photometer.

Reagents

Neutral normal ammonium acetate, potassium chloride standard solution

Special equipments

Flame photometer

Procedure

Transfer an accurately weighed quantity of soil (5 gm) to a shaking bottle, add 25 ml of neutral normal ammonium acetate (pH 7) and shake it for 5 minutes and filter immediately through a Whatman No 1. filter paper. Reject the first few ml of the filtrate and then collect the filtrate and determine the potassium present using a flame photometer. From the standard curve, find out the amount of potassium in the solution.

Calculations

Weight of soil taken = 5 gm

Volume of extractant added = 25 ml

Amount of potassium in the extract = X ppm

Amount of available K in kg/ha

$$= \text{ppm K} \times \frac{\text{Volume of the extractant}}{\text{Weight of soil}} \times 2.24 \times \frac{10^6}{10^6}$$

$$= \frac{X \times 25 \times 2.24}{5} \times \frac{10^6}{10^6}$$

$$= X \times 11.2 = \text{ppm K} \times 11.2$$

Classification

Low	< 113	kg/ha
Medium	113 - 280	kg/ha
High	> 280	kg/ha

ii. Nonexchangeable potassium

The nonexchangeable potassium is estimated by extracting the soil with strong inorganic acids like 6 N sulphuric acid, hot 1 N hydrochloric acid or boiling 1 N nitric acid and estimating the same flame photometrically.

22 SULPHUR (PHOSPHATE EXTRACTABLE SULPHUR) (Ensminger, 1954)

Principle

Number of extractants like phosphate solution, Morgan's reagent, Bray's, Olsen's normal magnesium acetate, 0.15% calcium chloride have been proposed for extracting sulphur in the soil. Of these, phosphate extractant is preferred by many workers. The extent of turbidity due to fine suspension of barium sulphate formed can be measured accurately in a nephlometer (working on the principle of scattering of light) and when present in low concentrations using spectrophotometer or even colorimeter.

Reagents

Potassium dihydrogen phosphate, nitric acid (25%), acetophosphoric acid, barium sulphate, barium chloride, gumacacia - acetic acid.

Special equipments

Colorimeter

Procedure

Transfer 20 gm of soil into a 250 ml conical flask, add 100 ml of 500 ppm of P solution prepared from potassium dihydrogen orthophosphate or monocalcium phosphate and shake it for 30 minutes. Filter the extract through Whatman No. 42 filter paper and determine the sulphate by turbidity.

The turbidity is estimated by the turbidimetric method for sulphate (Massount and Cornfield, 1963). Take an aliquot of the extract (not more than 15 ml containing less than 120 μg of sulphate - sulphur) in 25 ml volumetric flask and add 2.5 ml of 25% nitric acid and 2 ml of acetophosphoric acid and dilute it to 22 ml. Close it and shake. Then add 0.5 ml of barium sulphate seed suspension and 0.2 gm of barium chloride crystals and shake it thrice. After 10 minutes invert it again 10 times and after 5 minutes another 5 times. Again after 5 minutes, add 1 ml of gumacacia - acetic acid and dilute it to volume and invert three times and set aside for 90 minutes. Afterwards invert the flask again 10 times and read it in Klett - Summerson colorimeter at 440 mμ with blue filter or in a spectrophotometer. Using the standard curve, find out the concentration and from that calculate the amount of sulphate - sulphur present.

ii. Colorimetric method for available sulphate sulphur using barium chromate (Palaskar et al. 1981)

Principle

The sulphur is extracted with 0.15% calcium chloride in 1:5 soil : solution ratio using 10-20 gm soil and 30 minutes shaking time.

Reagents

Barium chromate (0.1 N in 1.5 N hydrochloric acid). ammonium hydroxide (approximately 5 N), standard sulphate solution (100 ppm S), calcium chloride (0.15%).

Procedure

Transfer an accurately weighed quantity of soil (10 gm) into a flask and add 50 ml of 0.15% calcium chloride and shake it for 30 minutes. Filter the solution through a dry filter paper.

Pipette out 5 ml of barium chromate solution into 100 ml volumetric flask, add 1 - 2 ml of 5 N ammonium hydroxide to reduce the free acidity to about 0.05 N. No turbidity should appear. Add 10 ml of the soil extract, shake it and keep it overnight. Then add 1 ml of ammonium hydroxide to precipitate the unreacted barium chromate and then make up to the volume with distilled water. Close the flask and shake it 2 to 3 times and filter the contents through dry Whatman No. 42 filter paper, rejecting the first few ml of the filtrate. Measure the intensity of yellow colour in a colorimeter using a deep blue filter (420 mμ).

The soil extract should have a pH around 2 to 3 before the addition of barium chromate. In case it is alkaline, adjust the pH by adding 6 N hydrochloric acid.

Calculations

Same as in other colorimetric estimations.

23. AVAILABLE ZINC (Lindsay and Norvell, 1969)

Principle

The available zinc in soil is extracted using 0.005 M DTPA (diethylene triamine pentaacetic acid) after shaking it for 2 hours. The amount of zinc extracted is estimated in an Atomic absorption spectrophotometer.

Reagents

DTPA (0.005 M), triethanolamine (TEA) (0.1 M), calcium chloride (0.01 M), zinc metal, dilute hydrochloric acid (concentrated acid diluted with double distilled water).

Special equipments

Atomic absorption spectrophotometer

Procedure

Transfer 10 gm of soil into a 100 ml conical flask and add 20 ml of DTPA extractant and shake well for 2 hours. Filter it using Whatman No. 40 or 42 filter paper and estimate the amount of zinc using Atomic absorption spectrophotometer. Using a standard curve find out the amount of zinc present in the extract and from this the amount present in the soil.

Calculations

Same as is given under colorimetric estimations.

24. LIME STATUS OF SOIL

Principle

The calcium carbonate in the soil sample is neutralised with excess of standard acid and the excess acid is back titrated with the standard alkali using phenolphthalein as indicator.

$CaCO_3 + H_2SO_4$ ------> $CaSO_4 + H_2O + CO_2$

$H_2SO_4 + 2\ KOH$ ------> $K_2SO_4 + 2\ H_2O$

Reagents

0.1 N sulphuric acid, phenolphthalein, 0.1 N potassium hydroxide.

Procedure

Transfer an accurately weighed amount of soil (5 gm) into a 250 ml beaker and add 100 ml of 0.1 N sulphuric acid, shake the suspension and allow it to stand for one hour to complete the reaction. Pipette 25 ml of clear supernatent solution to a 250 ml conical flask, add 2 drops of phenolphthalein and titrate against 0.1 N potassium hydroxide till the appearance of permanent pink colour.

Calculations

Weight of soil taken = W gm

Volume of 0.1 N sulphuric acid = 100 ml

Volume of extract taken for titration = 25 ml

Volume of 0.1 N potassium hydroxide consumed for 25 ml extract = V_1 ml

Volume of 0.1 N potassium hydroxide consumed for 25 ml of 0.1 N sulphuric acid = V_2 ml

Actual amount of 0.1 sulphuric acid used for neutralisation of calcium carbonate = $(V_2 - V_1)$ ml

1.0 ml of 0.1 N sulphuric acid = 0.005 gm of $CaCO_3$

$(V_2 - V_1)$ ml of 0.1 N sulphuric acid (25 ml extract)
= 0.005 x $(V_2 - V_1)$ gm $CaCO_3$

Hence in 100 ml of the extract (W gm soil)

$$= 0.005 \times (V_2 - V_1) \times \frac{100}{25}$$

Per cent of calcium carbonate

$$= 0.005 \times (V_2 - V_1) \times \frac{100}{25} \times \frac{100}{W}$$

Percent of calcium carbonate on moisture free basis

$$= 0.005 \times (V_2 - V_1) \times \frac{100}{25} \times \frac{100}{W} \times \frac{100}{(100 - M)}$$

25. LIME REQUIREMENT OF SOIL (Shoemaker's method, 1961)

Principle

This method is based on the depression of pH of a buffer solution when the soil is added to it. Lime requirement of a soil is proportional to the depression of pH of the buffer. For soils having pH 6.4 and above no lime is necessary.

Reagents

Extraction buffer adjusted to pH 7.5

Procedure

Transfer 5 gm of acid soil to a 100 ml beaker, add 5 ml of distilled water and 10 ml of extraction buffer solution. Shake the contents of the beaker for 10 minutes continuously or 20 minutes intermittently. Then measure the pH of soil - buffer suspension with a pH meter. From the pH of the soil suspension, find out the lime requirements using the table given below:

pH of soil suspension	Lime required to bring the soil to the pH of		
	6.0	6.4	6.8
	(calcium carbonate in tons/ha)		
6.7	2.47	2.96	3.46
6.6	3.46	4.20	4.69
6.5	4.45	5.43	6.18
6.4	5.68	6.67	7.66
6.3	6.67	7.90	9.14
6.2	7.66	9.14	10.37
6.1	8.65	10.37	11.86
6.0	9.63	11.61	13.34
5.9	10.87	12.84	14.82

5.8	11.86	14.08	16.06
5.7	12.84	15.31	17.54
5.6	13.83	16.55	19.02
5.5	14.82	17.78	20.50
5.4	16.06	19.02	21.98
5.3	17.04	20.54	23.22
5.2	18.28	21.24	24.70
5.1	19.27	22.48	26.18
5.0	20.54	23.71	27.66
4.9	21.24	24.95	29.15
4.8	22.48	26.18	30.63

26. GYPSUM REQUIREMENT (Schoonover's method)

Principle

A small quantity of soil when treated with a relatively large volume of standard, nearly saturated gypsum solution, will remove calcium from the solution and the loss of calcium gives a direct measure of the gypsum required to replace the sodium from the soil. The unreacted amount of gypsum is estimated by the EDTA titration. The method is based on the complexing of Ca^{2+}, Mg^{2+} and other ions by EDTA at different pH. The interference of ions like Fe^{2+} can be overcome by using sodium hydroxide (2%) solution or sodium carbonate (2%). A known volume of solution is titrated with 0.01 N versenate using murexide indicator in the presence of sodium hydroxide solution. The end point is change in colour from red to blue at pH 12 when all the calcium in solution forms a complex with EDTA.

Reagent

Calcium sulphate (saturated) solution, ammonium chloride - ammonium hydroxide buffer solution, erichrome black T indicator, EDTA (0.01N)

Procedure

Transfer 5 gm of soil into a 250 ml conical flask, add 100 ml of saturated calcium sulphate solution and shake for 5 minutes in a mechanical shaker and filter through Whatman No. 3 filter paper. Pipette out 5 ml of the filtrate into a porcelain basin, add 5 ml of buffer solution (add 1 ml of 2% sodium hydroxide if necessary) and 2 - 3 drops of erichrome black T indicator. Titrate with 0.01 N EDTA until the colour changes from red to sky blue. Run blank also using 5 ml of saturated calcium sulphate solution to determine the concentration of calcium originally present in it.

Calculations

Weight of soil = W gm

Volume of calcium sulphate added = 100 ml

Volume of EDTA equivalent to saturated calcium sulphate (blank) = V_1 ml

Volume of 0.01 N EDTA used for 5 ml of soil filtrate = V_2 ml

Volume of 0.01 N EDTA equivalent to calcium taken up by the soil = $(V_1 - V_2)$ ml

1 ml of 0.01 N EDTA = 0.0002 gm Ca

$(V_1 - V_2)$ ml of 0.01 N EDTA (= 5 ml of filtrate) $= 0.0002 \times (V_1 - V_2)$

100 ml of extract (= W gm of soil) $= 0.0002 \times (V_1 - V_2) \times \frac{100}{5}$

In 2 million kilograms of soil

$$= 0.0002 \times (V_1 - V_2) \times \frac{100}{5} \times 2 \times \frac{10^6}{5}$$

$$= 16 \times 10^2 \times (V_1 - V_2) \text{ kg/ha}$$

Gypsum requirement in tons/ha

$$= \frac{16 \times 10^2}{1000} \times (V_1 - V_2) \times \frac{172}{40}$$

If gypsum requirement has been calculated in meq/100 gm then the gypsum requirement in metric tons/ha for 30 cm soil depth = 1.72 x (meq of gypsum per 100 gm soil) x 2.47.

APPENDIX - I

INTERNATIONAL ATOMIC WEIGHTS

Name of element	Symbol	Atomic No.	Atomic Weight	Name of element	Symbol	Atomic No.	Atomic Weight
Hydrogen	H	1	1.008	Copper	Cu	29	63.54
Helium	He	2	4.003	Zinc	Zn	30	65.37
Lithium	Li	3	6.939	Gallium	Ga	31	69.72
Beryllium	Be	4	9.012	Germanium	Ge	32	72.59
Boron	B	5	10.811	Arsenic	As	33	74.922
Carbon	C	6	12.011	Selenium	Se	34	78.96
Nitrogen	N	7	14.007	Bromine	Br	35	79.909
Oxygen	O	8	15.999	Krypton	Kr	36	83.80
Fluorine	F	9	18.998	Rubidium	Rb	37	85.47
Neon	Ne	10	20.183	Strontium	Sr	38	87.62
Sodium (Natrium)	Na	11	22.99	Yttrium	Y	39	88.90
Magnesium	Mg	12	24.312	Zirconium	Zr	40	91.22
Aluminium	Al	13	26.982	Niobium (Coloumbium)	Cb	41	92.906
Silicon	Si	14	28.086	Molybdenum	Mo	42	95.94
Phosphorus	P	15	30.974	Technetium	Tc	43	98.906
Sulphur	S	16	32.064	Ruthenium	Ry	44	101.07
Chlorine	Cl	17	35.453	Rhodium	Rh	45	102.905
Argon	Ar	18	39.948	Palladium	Pd	46	106.4
Potassium (Kalium)	K	19	39.102	Silver (Argentium)	Ag	47	107.87
Calcium	Ca	20	40.08	Cadmium	Cd	48	112.40
Scandium	Sc	21	44.956	Indium	In	49	114.82
Titanium	Ti	22	47.90	Tin (Stannum)	Sn	50	118.69
Vanadium	V	23	50.942	Antimony (Stibium)	Sb	51	121.75
Chromium	Cr	24	51.996	Tellurium	Te	52	127.60
Manganese	Mn	25	54.938	Iodine	I	53	126.904
Iron (Ferrum)	Fe	26	55.847	Xenon	Xe	54	131.30
Cobalt	Co	27	58.933	Cesium	Cs	55	132.995
Nickel	Ni	28	58.71	Barium	Ba	56	137.34

Name of element	Symbol	Atomic No.	Atomic Weight	Name of element	Symbol	Atomic No.	Atomic Weight
Lanthanum	La	57	138.91	Thallium	Tl	81	204.37
Cerium	Ce	58	140.12	Lead (Lubum)	Pb	82	207.19
Praseodymium	Pr	59	140.907	Bismuth	Bi	83	208.980
Neodymum	Nd	60	144.24	Polonium	Po	84	209.98
Promethium	Pm	61	147.00	Astatine	At	85	209.99
Samarium	Sm	62	150.35	Radon	Rn	86	222.02
Europium	Eu	63	151.96	Francium	Fr	87	223.02
Gadolinium	Gd	64	157.25	Radium	Ra	88	226.03
Terbium	Tb	65	158.924	Actinium	Ac	89	227.03
Dysprosium	Dy	66	162.50	Thorium	Th	90	232.038
Holmium	Ho	67	164.930	Protoactinium	Pa	91	231.04
Erbium	Er	68	167.26	Uranium	U	92	238.03
Thulium	Tm	69	168.934	Neptunium	Np	93	237.05
Ytterbium	Yb	70	173.04	Plutonium	Pu	94	244.06
Intetium	In	71	174.97	Americum	Am	95	243.06
Hafnium	Hf	72	178.49	Curium	Cm	96	247.07
Tantalum	Ta	73	180.948	Berkelium	Bk	97	247.07
Tungsten (Wolfram)	W	74	183.85	Californium	Cf	98	251.08
Rhenium	Re	75	186.2	Einsteinium	Es	99	252.08
Osmium	Os	76	190.2	Fermium	Fm	100	257.10
Iridium	Ir	77	192.2	Mendelevium	Mi	101	258.10
Platinum	Pt	78	195.09	Nobelium	No	102	259.10
Gold (Aurum)	Au	79	196.967	Lawrencium	Lw	103	262.11
Mercury (Hydragyrum)	Hg	80	200.59	Dubnium	Db	104	261.11
				Jaliotium	Jl	105	262.11
				Rutherfordium	Rf	106	263.12
				Bohrium	Bh	107	262.12
				Hohnium	Hn	108	**
				Meiternium	Mt	109	**

APPENDIX II

IMPORTANT VOLUMETRIC PRIMARY STANDARDS

Primary standard	Formula	Eq. wt / Mol.wt	Eq. wt	Mol. wt
S- Diphenylguandine	$NH : C (NHC_6H_5)_2$	1	211.26	211.26
Meruric oxide	HgO	1/2	108.31	216.62
Potassium acid carbonate	$KHCO_3$	1	100.12	100.12
Potassium iodate	KIO_3	1/6	35.67	204.02
Sodium carbonate	Na_2CO_3	1/2	53.00	106.00
Sodium oxalate	$Na_2C_2O_4$	1/2	67.00	134.00
Sodium tetraborate (Borax)	$Na_2B_4O_7.10\ H_2O$	1/2	190.72	381.44
Benzoic acid	$C_6H_5\ COOH$	1	122.13	122.13
Hydrazine sulphate	$N_2H_4H_2\ SO_4$	1/2	65.07	130.14
Oxalic acid crystals	$H_2C_2O_4.\ 2\ H_2O$	1/2	63.03	126.06
Potassium acid oxalate	$KHC_2\ O_4$	1	128.13	128.13
Potassium acid phthalate	$KHC_8\ H_4\ O_4$	1	204.23	204.23
Potassium acid tartarate	$KHC_4\ H_4\ O_6$	1	188.18	188.18
Potassium tetraoxalate	$KH_3\ (C_2O_4)_2.\ 2H_2O$	1/3	84.73	254.20
Ferrous sulphate	$FeSO_4.\ 7\ H_2O$	1	278.03	278.03
Ferrous ammonium sulphate	$FeSO_4\ (NH_4)\ 2SO_4.\ 6H_2O$	1	392.16	392.16
Iron wire	Fe	1	55.85	55.85
Potassium ferrocyanide	$K_4\ Fe(CN)_6.\ 3H_2O$	1	422.41	422.41
Silver	Ag	1	107.88	107.88
Arsenious oxide	As_2O_3	1/4	49.455	197.82
Copper	Cu	1	63.54	63.54
Hydrazine sulphate	$N_2H_4\ .\ H_2\ SO_4$	1/4	32.531	130.124

Primary standard	Formula	Eq. wt / Mol.wt	Eq. wt	Mol. wt
Iodine (resublimed)	I	1	126.91	126.91
Iodine cyanide	ICN	1/2	76.465	152.93
Potassium bormate	$KBrO_3$	1/6	27.836	167.016
Potassium dichromate	$K_2 Cr_2 O_7$	1/6	49.037	294.222
Potassium ferricyanide	$K_3 Fe(CN)_6$	1	329.26	329.26
Sodium thiosulphate	$Na_2S_2O_3 . 5 H_2O$	1	248.21	248.21
Potassium bromide	KBr	1	119.02	119.02
Potassium chloride	KCl	1	74.56	74.56
Sodium chloride	NaCl	1	58.45	58.45
Mercury	Hg	1/2	100.31	200.62
Silver nitrate	$AgNO_3$	1	169.89	169.89

APPENDIX III

PREPARATION OF STANDARD CURVES FOR DIFFERENT ELEMENTS

In all instrumental analysis, first the instrument has to be standardised with an exactly known concentration of the element, prepared by dissolving an exact amount of the substance in a known volume of the solvent, usually water. If the substance is insoluble in water, then it is dissolved in other solvents, generally a dilute acid and made upto the desired volume. This forms the stock solution of the element. From this standard stock solution, solutions of various concentrations are prepared by suitable dilutions.

In case of colorimetric analysis, first a series of different intensities of colour is developed by adopting the concerned procedures and using different concentrations of the element from 0 to the maximum normally encountered in that particular analysis. The intensity of the colours thus developed is read in a colorimeter and the corresponding absorbance/ transmittance values are recorded. Then using these values, a curve is drawn on a graph paper having the concentrations on the X axis and the readings on the Y axis. All points are plotted on the graph and then they are connected either with a straight line or by smooth curve. If certain points fall far away from the general pattern, then the whole procedure is repeated to find out the defects in the estimation and rectify the same. After rectifying the same, the above estimations are repeated and the same procedure is followed and a smooth curve or line is drawn connecting all the points.

Once the standard curve is drawn for a particular element, this can be used to find out the concentration of the unknown solution. Adopting the same analytical procedure, the colour is developed using the test solution in the place of the standard solutions and the colorimetric readings recorded. Then the concentrations of the solution corresponding to the colorimetric reading is read from the standard curve. Invariably the concentrations in the standard curve is expressed in ppm. In case the same is required in per cent, it can be recalculated using the concerned equation.

It is absolutely necessary to prepare the standard curves for each instrument separately and not to use the standard curves prepared for some other instrument, even though both of them may be of the same type or even same brand. This is because there are always slight differences in the manufacture of the instruments resulting in variations in thickness of the tubes, path length of light, intensity of light, capacity of the photocell etc. Hence the standard curve prepared for one instrument cannot be used for another instrument.

Also for greater accuracy and precision, the standard curves should be prepared afresh when fresh standard solutions and reagents are used as there are liklihood of variations in the concentrations, amount of impurities or interfering radicals etc., and also when bulbs or any part of the instrument is changed. It is also advisable to prepare the standard curves at frequent intervals and also check the condition of the equipments by introducing the standard solutions either at random or at frequent intervals in each batch of samples that are being analysed each day.

If the estimations are made with a flame photometer, the standard solutions are fed into it and the readings are recorded and the standard curves prepared in a manner similar to that described for the colorimetric estimations of elements. The concentration of the test solutions is also found out in a similar manner.

The amount of the element to be taken for preparing stock solutions, nature of compounds to be used, concentrations of stock solutions normally prepared and the concentrations range normally used for preparing standard curves are given below:

Element	Compound to be used	Amount of salt (gm/volume)	Concentration of stock solution	Concentration used for preparing standard curve
P	KH_2PO_4	0.439 gm/1000 ml	100 ppm	2-20 ppm
K	KCl	1.907 gm/100 ml	1000 ppm	20-100 ppm
SO_4-S	K_2SO_4	1.080 gm/100 ml	2 mg-S/ml	10-120 ppm
Zn	Zn metal	1.000 gm/1000 ml	1000 ppm	0.5-2.5 ppm
Cu	Cu metal	0.100 gm/1000 ml	100 ppm	0.5-10 ppm
Biuret	Biuret	0.100 gm/100 ml	1000 ppm	0.1-5.0 ppm

The procedure for preparing the standard curves for different elements are briefly described below:

i. Phosphorus

Dry the potassium dihydrogen orthophosphate (KH_2PO_4) in an air oven at 50°C for one hour. Then accurately weigh 0.439 gm of this reagent in about 500 ml of distilled water, add 25 ml of approximately 7 N sulphuric acid and make it up to 1000 ml with distilled water. This gives 100 ppm stock standard solution of P (100 μg of P per ml). From this prepare a 2 μg solution of P (2 ppm) by taking a known volume of 100 ppm stock P solution and diluting it by 50 times.

For preparing the standard curve, take different volumes of 2 μg P solution (2 ppm) (1, 2, 3, 4, 5 and 10 ml) in 25 ml volumetric flasks. This will give 2, 4, 6, 8, 10 and 20 μg of P solutions. Add 5 ml of the extracting reagent (Bray/ Olsen) and develop the colour in all the six flasks by adding Dickman and Bray's reagent (I/II) and stannous chloride as given in the respective methods.

If the ascorbic acid method is followed, develop the colour after adjusting the pH of the aliquot.

Then take the colorimetric readings either transmittance or absorbance at 660 mμ (red filter) just after 10 minutes and plot the values in a graph paper with concentrations (or amount) of P on the X - axis and the colorimetric readings on the Y-axis.

In the case of other phosphorus analysis, prepare the standards as indicated above using the stock solution and add the concerned extractant or other reagents as in the procedures for the concerned estimations and develop the colour. Observe the readings of either transmittance or absorbance at 660 mμ (red filter). Then draw the standard curve as indicated above.

ii. Potassium

Dry pure potassium chloride (AR) in an air oven for 2 hours at 105°C. Then accurately weigh 1.907 gm of the dried potassium chloride and transfer it to 100 ml volumetric flask and make up to mark. This solution gives a stock solution of 1000 ppm K (1000 μg/ml).

Take 50 ml of that 1000 ppm K stock solution in a 100 ml volumetric flask and make up to the mark with water. This will give a stock solution of 500 ppm K.

Prepare a series of standard solutions containing 20, 40, 60, 80, and 100 ppm by taking suitable aliquots of the stock solution (500 ppm) (viz. 10, 20, 30, 40 and 50 ml respectively) into a 250 ml volumetric flask. Introduce the solutions in the flame photometer, observe the readings and plot them in a graph paper with concentration of K on the X axis and flame photometer readings on the Y axis. Connect them by a straight line or by a smooth curve and this gives a standard curve for K.

iii. Sulphate - sulphur (SO_4 -S)

Dry pure potassium sulphate (AR) in air oven for 2 hours at 105°C. Accurately weigh 1.080 gm of oven dried reagent and transfer it to 100 ml volumetric flask and dilute it to the mark. This gives a concentrated standard stock solution of 2 mg S/ ml (10 ppm).

Prepare a series of stock solutions by taking 1, 3, 5, 8, 10 and 12 ml of the working standard solution in 25 ml volumetric flask and adding the reagents as in the procedure for sulphate - sulphur. Plot the readings against the concentration on a graph paper and draw a curve.

If phosphate is present it is likely to interfere with the colour development. It can be precipitated out with ferric salt and ammonia during the same operation of removal of precipitated barium sulphate and chromate. If excessive amount of phosphate is present, turbidity method of estimation should be adopted.

iv. Zinc

Dissolve 1 gm of pure zinc metal in minimum amount of dilute hydrochloric acid and make upto 1000 ml with double distilled water and then transfer it to a plastic container. This stock solution contains 1000 μg Zn/ml (1000 ppm). From this prepare a working standard of 25 μg/ml by taking accurately 2.5 ml of the above stock solution and diluting it to 100 ml.

Prepare a series of standard solutions of 0.5, 1.0, 1.5, 2.0 and 2.5 ppm by transferring the concentrated aliquots of the working standard to a 100 ml volumetric flask and then adding the reagents as given in the procedure for estimation of zinc.

Plot on a graph paper the observed values against their respective concentrations of zinc and draw a curve connecting them. This gives a standard curve for zinc.

v. Biuret

Dissolve 100 mg of pure biuret reagent in carbon dioxide free water and transfer it to a 100 ml volumetric flask. Wash the beaker with minimum amount of water and transfer the washings to the flask and then make up to the mark. This gives a concentration of 1 mg/ml (1000 μg/ml or 1000 ppm).

Transfer different volumes of biuret standard solution into a 100 ml beaker and adjust the volume to approximately 50 ml with carbon dioxide free water. Add a drop of methyl red and neutralise with 0.1 N sulphuric acid to pink colour. Add 20 ml alkaline tartarate solution and then 20 ml of the copper sulphate solution. Then dilute it to volume, shake well for 10 seconds and place it in a water bath for 15 minutes at 30 ± 5°C. Prepare a blank and determine the absorbance of each solution against the blank at 555 mμ with 2-4 cm cell and prepare the standard curve as indicated in the earlier methods.

vi. Copper

Clean the surface of copper with sand paper. Dissolve 100 mg of pure copper in dilute nitric acid. Heat the solution nearly to dryness to drive away the acid. Add about 10 ml of water and heat again to near dryness. Take up the residue in water. Add 25 ml of 1 N sulphuric acid and dilute it to 1000 ml. One ml of this solution contains 0.1 mg of copper (as Cu).

Or alternatively dissolve 3.928 gm of cupric sulphate ($CuSO_4$. 5 H_2O) in sufficient amount of water containing one or two ml of concentrated sulphuric acid and dilute to 1000 ml. One ml of this solution will contain 1.0 mg of copper (as Cu) (1000 μg/ml) or 1000 ppm.

APPENDIX IV

IMPORTANT REAGENTS USED IN QUANTITATIVE ANALYSIS

1. ACETIC ACID (DILUTE)
2. ACETIC ACID (GLACIAL) (approximately 17.5 N)
3. ACETIC - PHOSPHORIC ACID: Mix 900 ml of glacial acetic acid (AR) with 300 ml of orthophosphoric acid
4. ACTIVATED CHARCOAL (DARCO G 60): Make it free of soluble P by repeated washings with 0.5 M sodium bicarbonate extractant adjusted to pH 8.5 (Olsen's reagent) and finally wash it with distilled water.
5. ALCOHOL - 35%
6. ALCOHOL (Methylated spirit) – 95-98%
7. ALKALINE PERMANGANATE SOLUTION (0.32%): Dissolve 3.2 gm of potassium permanganate in distilled water and make up to 1000 ml.
8. ALKALINE TARTARATE SOLUTION : Dissolve 40 gm of sodium hydroxide in 50 ml water, cool, add 50 gm of sodium potassium tartarate ($NaKC_4 H_4O_6 \cdot 4H_20$) and dilute it to 1000 ml. Allow it to stand for atleast 24 hours (for biuret estimation).
9. AMMONIUM ACETATE (NEUTRAL - NORMAL) - see neutral normal ammonium acetate.
10 AMMONIUM CARBONATE (saturated aqueous solution)
11. AMMONIUM CHLORIDE (solid)
12. AMMONIUM CHLORIDE - AMMONIUM HYDROXIDE BUFFER (pH 10): Dissolve 67.5 gm of pure ammonium chloride in 570 ml of concentrated ammonium hydroxide (precooled) and make up to 1000 ml with distilled water and adjust it to pH using a pH meter.

13. AMMONIUM CITRATE SOLUTION (40%): Dissolve 40 gm of ammonium citrate in 100 ml of water.
14. AMMONIUM CITRATE SOLUTION: Dissolve 370 gm of crystalline citric acid in 1.5 liters of distilled water and nearly neutralise by adding 345 ml of ammonium hydroxide (28 - 29% NH_3). If the concentration of ammonia is less than 28%, add correspondingly larger volume and dissolve citric acid in correspondingly smaller volume of water, cool and check the pH. Adjust with ammonium hydroxide (1:7) the pH of the citric acid solution to 7. Dilute the solution if necessary to 1.09 at 20°C. The total volume will be approximately 2 liters. Keep the solution in a tightly stoppered bottles and check pH from time to time. Adjust it to pH 7 if necessary (citrate insolubles estimation)
15. AMMONIUM HYDROXIDE (0.88 sp.gr.) Approximately 15 N.
16. AMMONIUM HYDROXIDE (1 : 4) (dilute): Dilute 1 part of ammonium hydroxide with 4 parts of water.
17. AMMONIUM HYDROXIDE (approximately 5 N): Dissolve 34 ml of ammonium hydroxide (0.88 sp.gr) in water and make up to 100 ml.
18. AMMONIUM HYDROXIDE (20%): Mix 20 ml of ammonium hydroxide with 80 ml of water.
19. AMMONIUM MOLYBDATE
20. AMMONIUM MOLYBDATE (20%): Dissolve 20 gm of ammonium molybdate in 100 ml of water, warm if necessary and cool.
21. AMMONIUM NITRATE (solid)
22. AMMONIUM OXALATE (solid)
23. AMMONIUM PURPURATE INDICATOR (MUREXIDE): Mix 0.5 gm of ammonium purpurate (murexide) thoroughly with 100 gm of potassium sulphate. Grind gently if necessary.
24. AMMONIUM THIOCYANATE (0.1 N): Dissolve 7.612 gm of ammonium thiocyanate (NH_4 SCN) in water and dilute it to 1000 ml. Then standardise this solution by titrating against 0.1 N silver nitrate solution using ferric alum as indicator.
25. ASCORBIC ACID SOLUTION : Dissolve 1.056 gm of ascorbic acid in 200 ml of reagent(A) (given below) and mix well. Prepare the solution fresh every time. (Reagent A;

Dissolve 12 gm of ammonium molybdate (AR) in 250 ml water. Also dissolve separately 0.291 gm of antimony tartarate in 100 ml of distilled water and add both these solutions to 1000 ml of 5 N sulphuric acid (140 ml of concentrated sulphuric acid diluted to 1000 ml). Mix thoroughly and make it up to 2000 ml with distilled water (for estimation of P by ascorbic acid method).

26. BARIUM CHLORIDE
27. BARIUM CHLORIDE (0.02 N): Dissolve exactly 2.0827 gm of barium chloride (AR) in 1000 ml of water.
28. BARIUM CHLORIDE (10%): Dissolve 100 gm of barium chloride in water, filter and make up to 1000 ml.
29. BARIUM CHROMATE (0.1 N in 1.5 N HCl): Dissolve 12.668 gm of barium chromate in 1.5 N hydrochloric acid and make it up to 1000 ml.
30. BARIUM SULPHATE SEED SUSPENSION: Dissolve 18.0 gm of barium chloride in 44 ml of hot water and add 0.5 ml of concentrated standard sulphate solution, boil it and cool it. Then add 4 ml of gumacacia - acetic acid solution. The seed suspension should be prepared every day (concentrated standard sulphate solution - 2 mg S /ml : Dissolve 1.080 of oven dried potassium sulphate in water and dilute to 100 ml).
31. BARTEN'S REAGENT: Dissolve 6.25 gm of ammonium molybdate in 100 ml warm distilled water and cool. In another container, dissolve 0.31 gm of ammonium metavanadate (AR) in 50 ml of hot water and add after cooling 62.5 ml of concentrated nitric acid (AR) and mix. Then add the ammonium molybdate solution to the ammonium metavanadate and make up to 250 ml. Mix the solution well and store in amber coloured bottles (for estimation of phosphorus). Or alternatively dissolve 25 gm of ammonium molybdate in 400 ml water separately. 1.25 gm of ammonium metavanadate in 300 ml concentrated nitric acid separately and then mix both the solutions and make up to 1000 ml.
32. BIURET STANDARD SOLUTION (1 mg/ml): Dissolve 100 mg of biuret (AR) in carbon dioxide free water and dilute to 100 ml.
33. BORIC ACID (2%): Dissolve 20 gm of boric acid in 1000 ml of water, warm if necessary.

34. BORAX STANDARD SOLUTION (100 ppm); Dissolve 0.572 gm of dried boric acid (H_3BO_3) in distilled water (not glass distilled water) and make up to 1000 ml. Use this for preparing working standards of 10 ppm. This can be done by diluting 1 part of the standard stock solution (100 ppm) with 9 parts of water.

35. BRAY'S No.1 REAGENT: Dissolve 2.775 gm of ammonium fluoride in 2500 ml of 0.025 N hydrochloric acid. This will give 0.03 N ammonium fluoride in 0.025 N hydrochloric acid.

36. BROMOCRESOL GREEN - METHYL RED INDICATOR; Dissolve 0.5 gm of bromocresol green and 0.1 gm of methyl red in 100 ml of 95% ethylalcohol and neutralise it to mid colour (bluish purple) with a few drops of N sodium hydroxide.

37. CALCIUM (0.02 N): Dissolve 1.001 gm of dried calcium carbonate (AR) in minimum quantity of dilute hydrochloric acid and make up to 1000 ml with water.

38. CALCIUM CHLORIDE (0.01 N): Dissolve 0.5 gm of calcium chloride in excess of dilute hydrochloric acid (10 ml) and make up to 1000 ml with distilled water.

39. CALCIUM OXIDE (Freshly ground)

40. CALCIUM SULPHATE SOLUTION (Saturated): Dissolve 5 gm of pure calcium sulphate ($CaSO_4 . 2 H_2O$) in 1000 ml of water, shake well for 10 minutes and filter.

41. CARBON TETRACHLORIDE

42. CATALYST MIXTURE; Mix potassium sulphate, copper sulphate and mercuric oxide in 10:1:0.1 ratio after grinding them separately (for phosphorus estimation).

43. CHLOROMOLYBDIC ACID REAGENT (Dickman-Bray's reagent having excess acid): Dissolve 1.5 gm of ammonium molybdate (AR) in 300 ml of distilled water (60°C), cool it and filter if necessary. To this add 400 ml of 10 N hydrochloric acid and make up to 1000 ml (for estimation of P by Olsen's method).

44. CITRIC ACID

45. CITRIC - MOLYBDIC ACID REAGENT: Dissolve 54 gm of 100% molybdic anhydride (MoO_3) and 12 gm of sodium hydroxide with stirring in 400 ml of hot water and cool. Dissolve 60 gm of citric acid in a mixture of 140 ml hydrochloric acid and 300 ml water and cool. Gradually add

molybdic solution to citric acid solution with stirring - cool and filter and dilute to 1000 ml (solution may be green or blue depending upon the exposure to light). If necessary add 0.5% $KBrO_3$ (potassium bromate) solution drop by drop until green colour becomes pale. Store in a dark place in polythene container (for use in quinoline method)

46. COBALTINITRITE (20%): Dissolve 200 gm of cobaltinitrite in 1000 ml of water.

47. COCHINEAL: Digest 3 gm of pulverised cochineal in a mixture of 50 ml of 95% alcohol and 200 ml water for one or two days at room temperature with frequent shaking, then filter and store.

48. COPPER SULPHATE

49. COPPER SULPHATE SOLUTION : Dissolve 15 gm of copper sulphate ($CuSO_4$. 5 H_2O) in carbon dioxide free water and dilute in 1000 ml.

50. CLAYTON YELLOW INDICATOR (0.04%): Dissolve 40 mg of clayton yellow in 100 ml of water (for sodium tetraphenyl boron method)

51. CURCUMIN - OXALIC ACID : Dissolve 0.04 gm of finely ground pure curcumin and 5 gm of oxalic acid (AR) in 95% ethyl alcohol and store it in amber coloured bottles in a cool dark place. Fresh reagents may be prepared after every few days (for boron estimation).

52. DEVARDA'S ALLOY

53. DICKMAN AND BRAY'S REAGENT: Dissolve 15 gm of ammonium molybdate in 300 ml of distilled water, warm to 60°C, filter if necessary after cooling. Then add 300 ml of 10 N hydrochloric acid and make up to 1000 ml. The normality of hydrochloric acid should be adjusted correctly by titration.

54. DIETHYLENE TRIAMINE PENTAACETIC ACID - see DTPA.

55. DIPHENYLAMINE INDICATOR : Dissolve 0.5 gm of diphenylamine in a mixture of 100 ml of concentrated sulphuric acid and 20 ml of distilled water containing 1.25% silver sulphate. If the soils or materials are free of chloride, addition of silver sulphate may be omitted.

56. DOUBLE INDICATOR : Dissolve 0.5 gm of bromocresol green and 0.1 gm of methyl red in 100 ml alcohol.

57. DTPA (Diphenyl triamine pentaacetic acid) (0.005 M) : Dissolve 1.967 gm of DTPA and 1.470 gm of calcium chloride

($CaCl_2 . H_2O$) in 20 - 25 ml of distilled water. Then add 138 ml of triethanolamine (0.1M) followed by 100 ml of water. Transfer it to 1000 ml volumetric flask and make up to volume with water. Adjust to pH 7.3 with dilute hydrochloric acid (for estimation of available zinc).

58. EDTA {Ethylene diamine tetraacetic acid (0.01 N)}: Dissolve 2.0 gm of EDTA disodium salt in distilled water and add 0.05 gm magnesium chloride hexahydrate ($MgCl_2 . 6 H_2O$). Make up to 1000 ml and shake well. Standardise this against 0.01 N calcium chloride (This is prepared by accurately weighing 0.50 gm of oven dried calcium carbonate and dissolving it in minimum excess of dilute hydrochloric acid and making it to 1000 ml).

59. EDTA (0.02 N): Weigh accurately 4.0 gm of EDTA sodium salt and dissolve it in distilled water. Add 0.1 gm of magnesium chloride hexahydrate ($MgCl_2 . 6 H_2O$) and make up to 1000 ml and standardise as before.

60. ERICHROME BLACK T INDICATOR (Solochrome black): Dissolve 0.5 gm of erichrome black T and 4.5 gm of hydroxylamine hydrochloride in 100 ml of 95% ethyl alcohol.

61. ETHYL ALCOHOL (sp.gr. 0.79)

62. ETHYLENE DIAMINE TETRAACETIC ACID (EDTA); see EDTA.

63. EXTRACTION BUFFER: Dissolve 1.8 gm of pure nitrophenol, 2.0 gm of calcium acetate, 2.5 gm of triethanolamine, 3.0 gm of potassium chromate and 53.1 gm of calcium chloride in 1000 ml of distilled water and adjust to pH 7.5 with sodium hydroxide.

64. FERRIC AMMONIUM SULPHATE: Prepare a cold saturated solution of ferric ammonium sulphate in water (approximately 40%) and add 1000 ml of concentrated nitric acid. One ml of the solution may be used for each titration.

65. FERROUS AMMONIUM SULPHATE (MOHRS SALT)

66. FERROUS AMMONIUM SULPHATE (0.5 N) : Dissolve 392.0 gm of ferrous ammonium sulphate {$Fe(NH_4)_2$ SO_4} 6 H_2O} in distilled water. Add 15 ml concentrated sulphuric acid and make up to 2000 ml with distilled water.

67. FORMALDEHYDE (37%)

68. FORMALDEHYDE SOLUTION: Dissolve one volume of formaldehyde with eight volumes of water.

69. GUMACACIA - ACETIC ACID SOLUTION; Dissolve 5 gm of chemically pure gum acacia in 500 ml hot water and filter the hot solution through Whatman No.42 filter paper. Cool the filtrate and dilute to 1000 ml with acetic acid.

70. HYDROCHLORIC ACID (concentrated): approximately 12 N.

71. HYDROCHLORIC ACID (Dilute): Dilute 240 ml of concentrated hydrochloric acid to 1000 ml with water.

72. HYDROCHLORIC ACID (1.0 N): Take 84 ml of concentrated hydrochloric acid and dilute it to 1000 ml with distilled water. Standardise it against a primary standard.

73. HYDROCHLORIC ACID (0.5 N) : Take 500 ml of 1.0 N hydrochloric acid and dilute to 1000 ml with water.

74. HYDROXYLAMINE HYDROCHLORIDE

75. INDICATOR SOLUTION : Mix three volumes of thymolblue solution and two volumes of phenolphthalein solution (Thymol blue solution: Dissolve 250 mg of thymolblue in 5.5 ml of 0.1 N sodium hydroxide and 125 ml of rectified spirit, dilute with water to 1250 ml. Phenolphthalein solution: Dissolve 250 mg of phenolphthalein in 100-150 ml of methylated spirit and dilute it with water to 250 ml).

76. IRON POWDER (reduced)

77. MAGNESIUM CHLORIDE (0.02 N) : As magnesium chloride is highly hygroscopic, the required quantity of magnesium metal (AR) or magnesium carbonate may be dissolved in little excess of dilute hydrochloric acid and then made up to 1000 ml.

78. METHYL ORANGE (0.5%) : This indicator is available either as a free acid or as its sodium salt. Dissolve 0.5 gm of the free acid in 1000 ml of water and filter the cold solution if necessary. Or dissolve 0.5 gm of sodium salt in 1000 ml of water and add 15.2 ml of 0.1 N hydrochloric acid. Filter if necessary in cold.

79. METHYL RED (0.5%): Dissolve 0.5 gm of methyl red in 100 ml of alcohol

80 METHYL RED (0.15%): Dissolve 0.15 gm of methyl red in 100 ml alcohol.

81. METHYL RED - METHYLENE BLUE MIXED INDICATOR: Mix equal volumes of 0.2% solution of methyl red in rectified spirit and 0.1% solution of methylene blue in rectified spirit.

82. MIXED INDICATOR: Dissolve 0.066 gm of methyl red with 0.099 gm of bromocresol green in 100 ml of 95% alcohol.

84. NEUTRAL NORMAL AMMONIUM ACETATE: Dissolve 77.09 gm of solid ammonium acetate in 20 liters of water and adjust it to pH 7.0 by addition of either ammonia or acetic acid.

84. NITRIC ACID (concentrated) : approximately 15.6 N.

85. NITRIC ACID (0.1619 N) : Add 11 ml of concentrated nitric acid to small amount of water and dilute it to 1000 ml. Standardise it against a primary standard.

86. OLSEN'S REAGENT: Dissolve 42.0 gm of sodium bicarbonate in 1000 ml of distilled water and adjust the pH to 8.5 using sodium hydroxide (20%).

87. ORTHOPHENONTHROLINE MONOHYDRATE: Dissolve 1.485 gm of orthophenonthroline monohydrate in 100 ml of 0.025 M ferrous sulphate (0.6938 gm of ferrous sulphate dissolved in 100 ml water).

88. ORTHOPHOSPHORIC ACID (85%) : approximately 15 N.

89. OXALIC ACID (0.1 N) : Dissolve 6.303 gm of oxalic acid predried in an air oven at 105 - 110°C in 1000 ml of water and standardise it against standard permanganate.

90. PERCHLORIC ACID (60%) (sp.gr 1.53), approximately 9 N: (sp.gr- 1.68; 72%) approximately 12 N.

91. PERCHLORIC ACID (20%)

92. PHENOLDISULPHONIC ACID : To 25 gm of pure phenol (C_6H_5OH - white crystals) kept in a dry conical flask (50 ml) add 150 ml of concentrated sulphuric acid and 75 ml of nitrate free fuming sulphuric acid and mix very carefully. Keep the contents on boiling water bath for 2 hours, covered with watch glass. After cooling, the phenoldisulphonic acid (in excess sulphuric acid) is stored in amber coloured bottles. In case fuming sulphuric acid is not available, the same quantity of concentrated sulphuric acid may be added instead and kept on boiling water bath for at least 6 hours.

93. PHENOLPHTHALEIN (0.5%): Dissolve 0.5 gm of phenolphthalein in 100 ml of 60% ethyl alcohol.

94. POTASSIUM CHLORIDE (1.0 N): Dissolve 74.0 gm of potassium chloride in water and make up to 1000 ml. For standard solutions of other lower normalities, dilute this solution suitably.

95. POTASSIUM CHROMATE INDICATOR: Dissolve 5 gm of potassium chromate in 100 ml water.
96. POTASSIUM CHROMATE (10%): Dissolve 10 gm of potassium chromate in 100 ml water. The indicator solutions should be free of chloride and should be tested by adding silver nitrate and observing whether a red precipitate appears or not. Filter the solution before use.
97. POTASSIUM DICHROMATE (1 N): Dissolve 49.04 gm of potassium dichromate in 1000 ml of water.
98. POTASSIUM FERRICYANIDE SOLUTION (1%)
99. POTASSIUM HYDROXIDE (0.1 N): Dissolve 5.7 gm of potassium hydroxide in 1000 ml of water and standardise against a standard acid.
100. POTASSIUM HYDROXIDE (0.1619 N): Dissolve 9.3 gm of potassium hydroxide in 1000 ml of water and standardise against an acid.
101. POTASSIUM HYDROXIDE (40%): Dissolve 40 gm of potassium hydroxide in 100 ml of water.
102. POTASSIUM PERMANGANATE (0.1 N): Dissolve 3.3 gm of potassium permanganate in 1000 ml of water and keep it overnight. The solution should not come in contact with any organic matter. After keeping it overnight, filter if necessary, and standardise it against 0.1 N oxalic acid or sodium oxalate.
103. POTASSIUM SULPHATE
104. POTASSIUM THIOCYANATE (1%): Dissolve 1 gm of potassium thiocyanate in 100 ml of water.
105. PRECIPITATING MIXTURE: Mix 10 ml of 20% ammonium molybdate with 7 ml of concentrated nitric acid and 3 ml of water. The solution should be clear. If there is any turbidity prepare the solution afresh.
106. QUATERNARY AMMONIUM CHLORIDE SOLUTION (approximately 0.625%): Dilute 50 ml of 12.8 zephiran chloride with water to 1000 ml, mix well and standardise. Cetyl trimethyl ammonium bromide may be substituted for Zephiran chloride. If other concentrations are used, adjust the volume.
107. QUIMOCIAC REAGENT: Dissolve 70 gm of sodium molybdate dihydrate in 150 ml water. Dissolve 60 gm of citric acid in a mixture of 80 ml nitric acid and 150 ml of

water and cool. Gradually add molybdate solution to citric acid - nitric acid mixture with stirring. Dissolve 5 ml of synthetic quinoline in mixture of 35 ml of nitric acid and 100 ml water. Gradually add this solution to molybdate - citric-nitric acid solution, mix and let it stand for 24 hours. Filter, add 28 ml acetone, dilute to 1000 ml with water and mix well. Store in polythene bottle.

108. SALICYLIC ACID

109. SILVER NITRATE (0.1 N): Dissolve 17.0 gm of dry silver nitrate in 1000 ml distilled water and standardise it against standard sodium chloride. Store it in amber coloured bottles.

110. SILVER NITRATE (0.01 N): Dissolve 1.70 gm of dry silver nitrate in 1000 ml distilled water and standardise it.

111. SILVER NITRATE (0.02 N): Dissolve 3.4 gm of dry silver nitrate in 1000 ml of distilled water and standardise. Or alternatively dilute the silver nitrate (0.1 N) solution suitably and use it.

112. SODIUM CHLORIDE (Saturated): Dissolve about 350 gm of sodium chloride (AR) in 1000 ml and maintain a small amount of salt at the bottom for some time. Then filter and use it.

113. SODIUM CHLORIDE (0.02 N) : Dissolve 1.170 gm of sodium chloride (AR) (predried at 80°C for 1 hour) in double distilled water and make it up to 1000 ml.

114. SODIUM CYANIDE

115. SODIUM DIETHYL DITHIOCARBAMATE SOLUTION (0.1%): Dissolve 1 gm of sodium diethyl dithiocarbamate in 1000 ml water.

116. SODIUM FLUORIDE

117. SODIUM HYDROXIDE (5 N): Dissolve 200 gm of pure sodium hydroxide in water and make it up to 1000 ml and standardise it against standard acid.

118. SODIUM HYDROXIDE (0.5 N): Dissolve 20 gm of pure sodium hydroxide in water and make up to 1000 ml. Standardise it against a standard acid. Or alternatively dilute 100 ml of exactly 5 N sodium hydroxide to 1000 ml with distilled water.

119. SODIUM HYDROXIDE (0.1 N): Dissolve 4 gm of pure sodium hydroxide in water and make up to 1000 ml. Standardise it against a standard acid. Or alternatively

dilute 20 ml of exactly 5 N sodium hydroxide to 1000 ml with distiled water.

120. SODIUM HYDROXIDE (0.2 N): Dilute 200 ml of 1 N sodium hydroxide to 1000 ml with water.

121. SODIUM HYDROXIDE (40%): Dissolve 40 gm of sodium hydroxide in 100 ml of distilled water.

122. SODIUM HYDROXIDE (20%): Dissolve 20 gm of sodium hydroxide in 100 ml of distilled water.

123. SODIUM NITRITE (35%): Dissolve 35 gm of sodium nitrite in 100 ml of distilled water.

124. SODIUM TETRAPHENYLBORON SOLUTION (approximately 12%) : Dissolve 12 gm of sodium tetraphenyl boron in approximately 800 ml of water. Add 20 - 25 ml of aluminium hydroxide, stir for 5 minutes, filter through Whatman No. 42 filter paper into a volumetric flask. Rinse the beaker sparingly with water and add to the filtrate. Collect the entire filtrate, add 2 ml of 20% sodium hydroxide solution, dilute to volume with water and mix. Let it stand for 48 hours and standardise. Adjust it so that 1 ml of sodium tetraphenyl boron solution is equal to 1% K_2O. Store at room temperature.

125. STANNOUS CHLORIDE: Dissolve 10 gm of crystalline stannous chloride (LR) in 25 ml of concentrated hydrochloric acid by warming and store it in an amber coloured bottle. This is 40% stannous chloride stock solution. Just before use dilute 0.5 ml of this stock solution with 86 ml distilled water. A piece of tin metal (AR) may be added to keep the solution for a longer time (for zinc estimation, use 0.1 ml instead of 0.5 ml of the stock solution)

126. STARCH INDICATOR: Stir one gram of soluble starch in 5 ml of cold water in a beaker. Pour the resulting suspension with constant stirring into a 100 ml of boiling water, cool it and transfer it to a 100 ml bottle. (A fairly stable starch is prepared by dissolving about 2 gm of boric acid in 100 ml of solution prepared as described above). Add a small quantity of preservative like chloroform or a few drops of 5% sodium hydroxide or 10 ml of 1% salicylic acid per liter of the starch solution or a few ml of 10% zinc chloride solution. A substitute for starch is Thyodene. It is a white powder and is fairly stable. Add 0.5 gm per titration.

127. SURFACE ACTIVE AGENT (0.5%): Dissolve 0.5 gm of sodium dodecyl benzene sulphonate in 100 ml of water.

128. SULPHURIC ACID (concentrated), approximately 36 N.

129. SULPHURIC ACID (0.1 N): Add 3.0 ml of concentrated sulphuric acid (36 N) to water and dilute it 1000 ml and then standardise it against primary standard.

130. SULPHURIC ACID (0.01 N): Add 1.0 ml of concentrated sulphuric acid (36 N) to water and dilute it 3600 ml and then standardise it against primary standard.

131. SUCROSE

132. THIOSULPHATE SOLUTION: Dissolve 80 gm of sodium thiosulphate ($Na_2\ S_20_3$. 5 H_20) in water and make it up to 1000 ml.

133. TRIETHANOLAMINE

134. TRIPLE ACID: Mix concentrated nitric acid : concentrated sulphuric acid: perchloric acid in the ratio 9:2:1. Use it immediately.

135. UREASE (Neutral solution): Shake 1 gm of jack bone meal with 100 ml of water for 5 minutes. Transfer 10 ml of the solution to 250 ml conical flask, dilute with 50 ml water, add 4 drops of methyl purple indicator. Titrate with 0.1 N hydrochloric acid to reddish purple, then back titrate to green with 0.1 N sodium hydroxide. From the difference in volume in ml. calculate the amount of 0.1 N hydrochloric acid required to neutralise the remainder of the solution (usually 2.5 ml/100ml). Add this amount and shake well.

136. WASH SOLUTION: Add potassium perchlorate to alcohol and shake until saturated solution is obtained. Keep the solution over a solid potassium perchlorate and filter immediately before use.

137. ZEPHIRAN CHLORIDE : To 1 ml of sodium tetraphenylboron solution in 125 ml conical flask, add 20 - 25 ml of water, 1 ml of 20% sodium hydroxide, 2.5 ml of formal dehyde, 1.5 ml of 4% ammonium oxalate {$(NH_4)_2\ C_2O_4$} and 6 - 8 drops of clayton yellow indicator. Titrate to pink end point with zephiran chloride solution using 10 ml microburette. Adjust zephiranchloride solution so that 2.0 ml of it is equal to 1 ml of sodium tetraphenyl boron solution.

138. ZINC DUST

139. ZINC STANDARD SOLUTION: Weigh 1.25 gm of reagent grade zinc metal and dissolve in 20 ml hydrochloric acid (1:1) and make it up to 500 ml.

APPENDIX V

IMPORTANT GRAVIMETRIC CONVERSION FACTORS

Multiply the weight of	By this factor	To obtain weight of
Al	1.8895	Al_2O_3
Al_2O_3	0.52913	Al
Ba	1.6994	$BaSO_4$
$BaSO_4$	0.58845	Ba
$BaSO_4$	0.13737	S
$BaSO_4$	0.41154	SO_4
Ca	2.4973	$CaSO_4$
Ca	4.2959	$CaSO_4 . 2 H_2O$
$CaCO_3$	0.4004	Ca
$CaSO_4 . 2 H_2O$	0.23277	Ca
CaO	0.715	Ca
CaC_2O_4 (oxalate)	0.3128	Ca
$CaSO_4 . 2 H_2O$	0.5579	SO_4
$CaCO_3$	0.43971	CO_2
AgCl	0.24737	Cl
Cl	4.0426	AgCl
KCl	0.47557	Cl
Cl	2.1027	KCl
NaCl	0.60664	Cl
Cl	1.6484	NaCl
Cu	3.9296	$CuSO_4 . 2 H_2O$
$CuSO_4 . 5 H_2O$	0.25448	Cu
$FeSO_4 (NH_4)_2 SO_4 . 6 H_2O$	0.14242	Fe
Fe	1.4297	Fe_2O_3
Fe_2O_3	0.69944	Fe
$FeSO_4 . 7 H_2O$	0.20088	Fe
$FeSO_4 . 7 H_2O$	0.28720	Fe_2O_3
MgO	0.60317	Mg
$Mg_2 P_2O_7$	0.2185	Mg
$MgSO_4 . 7 H_2O$	0.09866	Mg
MgP_2O_7	0.36228	MgO
N	1.2159	NH_3
N	4.7168	$(NH_4)_2 SO_4$

N	4.4266	NO_3
NH_3	0.82245	N
NH_3	3.6407	NO_3
$NaNO_3$	0.1648	N
$(NH_4)_2 SO_4$	0.21201	N
NO_3	0.22591	N
NO_3	0.27467	NH_3
NO_2	0.30447	N
P	2.2914	P_2O_5
P	3.0661	PO_4
PO_4	0.32613	P
PO_4	0.7473	P_2O_5
P_2O_5	0.43642	P
P_2O_5	2.18	$Ca\ (PO_4)_2$
P_2O_5	1.3381	PO_4
$Ca\ (PO_4)_2$	0.45762	P_2O_5
$Ca\ (PO_4)_2$	0.1997	P
$Ca\ (H_2PO_4).\ 2\ H_2O$	0.2457	P
$KH_2\ PO_4$	0.2276	P
K	1.2046	K_2O
KCl	0.52443	K
K_2O	0.83015	K
$K_2\ SO_4$	0.44874	K
$KClO_4$	0.28219	K
$AgCl$	0.2473	Cl
$AgNO_3$	0.84371	$AgCl$
$AgNO_3$	0.2087	Cl
Na	1.5422	Cl
Na_2CO_3	1.2855	$CaSO_4$
$NaCl$	0.6066	Cl
$NaCl$	0.39336	Na
$NaHCO_3$	0.27367	Na
$Na_2\ SO_4$	0.32371	Na
S	5.3695	$CaSO_4 .\ 2\ H_2O$
S	7.2795	$BaSO_4$
S	2,995	SO_4
$BaSO_4$	0.13737	S
$BaSO_4$	0.41155	SO_4
$CaSO_4 .\ 2\ H_2O$	0.1862	S
$CaSO_4 .\ 2\ H_2O$	0.5566	SO_4
SO_4	1.7812	$CaSO_4 .\ 2H_2O$
SO_4	0.3217	S
ZnO	0.8033	Zn
Zn	1.2447	ZnO
$ZnSO_4 .\ 7\ H_2O$	0.22736	Zn
Zn	4.3982	$ZnSO_4 .\ 7\ H_2O$

APPENDIX VI

PROPERTIES OF WHATMAN QUANTITATIVE FILTER PAPERS (Whole numbers give the filter paper number)

Filter paper	Fast speed retains coarse particles	Medium speed retains medium sized particles	Slow speed retains fine particles
Double acid washed-hardened	541	540	542
Ash (mg)	0.08	0.08	0.09

Filter	Qualitative	Qualitative wet strengthed	Ash less	Hardened	Hardened ash less	Retention
Fast	4	114,113	41	54	541	Coarse and gelatinous precipitate
Medium fast	1	111	43	—		Medium crystalline
Medium	2	—	40	52	540	Crystalline
Slow	5.6	—	44, 42	50	542	Fine crystalline
Ash (%)	0.06	0.06	0.01	0.025	0.008	

3 - Thick medium speed paper with high retention
GF/A - Standard laboratory glass microfibre paper
GF/B - Thicker stronger version of GF/A
GF/C - Very fine glass microfibre paper
GF/D - Thick version, ideal for membrane refiltration
GF/F - Ultrafine filter with retention down to 0.7 μm in liquid

APPENDIX VII

NUTRIENT CONTENT OF SOME IMPORTANT FERTILIZERS AND MANURES
(in per cent)

	Fertilizer/ manure	N	P_2O_5	K_2O
1	Ammonium sulphate	20.6	—	—
2	Ammonium sulphate nitrate	26.0	—	—
	(NH_3 - N = 19.5%: NO_3 - N = 6.5%)			
3	Ammonia - anhydrous	82.2	—	—
4	Urea	46.2	—	—
5	Ammonium chloride	26.0	—	—
6	Sodium nitrate	16.0	—	—
7	Calcium ammonium nitrate	20.5	—	—
	(NH_4 - N =10.5%; NO_3 - N = 10.5%)			
8	Superphosphate			
	Water soluble P_2O_5	—	16.0	
	Total P_2O_5	—	18.0	
9	Triple superphosphate	—	40.0	
10	Bone meal (raw) - total	3.0	20.0	—
	- citrate soluble	—	8.0	—
11	Bone meal (steamed) - total	—	22.0	—
	- citrate soluble	—	16.0	
12	Rockphosphate - total P_2O_5	—	25.0 to 39.0	—
13	Potassium chloride	—	—	60.0
14	Potassium sulphate	—	—	52.0
15	Monoammonium phosphate	11.0	40.0	—
16	Diammonium phosphate	21.0	53.0	—
	MANURES*			
17	Farm yard manure	1.2	0.6	1.2
18	Sheep manure	1.9	1.3	2.3
19	Poultry manure	0.9	1.8	0.6
20	Bates and guano manures	8.0	2.6	1.4
21	Pig manure	3.7	3.3	0.4
22	Compost - municipal	0.4	0.6	0.4
23	Compost leaf	1.0	0.7	1.0

PERIODIC TABLE OF THE ELEMENTS

PERIOD

KEY

ATOMIC NUMBER (4)	16
ATOMIC WEIGHT (2)	32.06
OXIDATION STATES (7)	6, 3, 4, −2
BOILING POINT °C (5)	444.6
MELTING POINT °C (6)	119.0
DENSITY (g/ml) (3)	2.07
SYMBOL (1)	S
ELECTRON STRUCTURE	$[Ne]3s^2 3p^4$
NAME	Sulfur

PERIODIC TABLE OF THE ELEMENTS

Period	Group	Atomic number	Symbol	Name	Atomic weight	Oxidation states	Boiling point °C	Melting point °C	Density (g/ml)	Electron structure
1	IA	1	H	Hydrogen	1.0079	±1	−252.7	−259.2	0.071	s^1
1	INERT GASES	2	He	Helium	4.0026	0	−268.9	−269.7	0.126	s^2
2	IA	3	Li	Lithium	6.941	1	1330	108.5	0.53	$s^2 2s^1$
2	IIA	4	Be	Beryllium	9.01213	2	2770	1277	1.85	$s^2 2s^2$
2	IIIA	5	B	Boron	10.811	3	—	~2200	2.34	$s^2 2s^2 2p^1$
2	IVA	6	C	Carbon	12.011	±4 2	4830	3727	2.26	$s^2 2s^2 2p^2$
2	VA	7	N	Nitrogen	14.0067	±3,5,4,2	−195.8	−210	0.81	$s^2 2s^2 2p^3$
2	VIA	8	O	Oxygen	15.9994	−2	−183	−218.8	1.14	$s^2 2s^2 2p^4$
2	VIIA	9	F	Fluorine	18.9984	−1	−188.2	−219.6	1.11	$s^2 2s^2 2p^5$
2	INERT GASES	10	Ne	Neon	20.179	0	−246	−248.6	1.20	
3	IA	11	Na	Sodium	22.98977	1	892	97.8	0.97	$[Ne]3s^1$
3	IIA	12	Mg	Magnesium	24.305	2	1107	650	1.74	$[Ne]3s^2$
3	IIIA	13	Al	Aluminium	26.98154	3	2450	660	2.70	$[Ne]3s^2 3p^1$
3	IVA	14	Si	Silicon	28.0855	4	2680	1410	2.33	$[Ne]3s^2 3p^2$
3	VA	15	P	Phosphorous	30.97376	± 3,5,4	280w	44.2w	1.82w	$[Ne]3s^2 3p^3$
3	VIA	16	S	Sulfur	32.06	6,3,4,-2	444.6	119.0	2.07	$[Ne]3s^2 3p^4$
3	VIIA	17	Cl	Chlorine	35.453	±1,4,5,6,7	−34.7	−101.0		$[Ne]3s^2 3p^5$
3	INERT GASES	18	Ar	Argon	39.948	0	−185.8	−189.4	1.40	$[Ne]3s^2 3p^6$
4	IA	19	K	Potassium	39.0983	1	760	63.7	0.86	$[Ar]4s^1$
4	IIA	20	Ca	Calcium	40.08	2	1440	838	1.55	$[Ar]4s^2$
4	IIIB	21	Sc	Scandium	44.9559	3	2730	1539	3.0	$[Ar]3d^1 4s^2$
4	IVB	22	Ti	Titanium	47.88	4,3	3260	1668	4.51	
4	VB	23	V	Vanadium	50.9415	5,4,3,2	3450	1917	6.1	$[Ar]3d^3 4s^2$
4	VIB	24	Cr	Chromium	51.996	6,3,2	2665	1875	7.19	$[Ar]3d^5 4s^1$
4	VIIB	25	Mn	Manganese	54.938	7,6,4,2,3	1245	7.43	7.19	$[Ar]3d^5 4s^2$
4	VIII B	26	Fe	Iron	55.847	2,3	3000	1536	7.86	$[Ar]3d^6 4s^2$
4	VIII B	27	Co	Cobalt	58.9332	2,3	2900	1495	8.9	$[Ar]3d^7 4s^2$
4	VIII B	28	Ni	Nickel	58.69	2,3	2730	1453	8.9	$[Ar]3d^8 4s^2$
4	IB	29	Cu	Copper	63.546	2,1	2595	1083	8.96	$[Ar]3d^{10} 4s^1$
4	IIB	30	Zn	Zinc	65.38	2	906	419.5	7.14	$[Ar]3d^{10} 4s^2$
4	IIIA	31	Ga	Gallium	69.72	3	2237	29.8	5.91	$[Ar]3d^{10} 4s^2 4p^1$
4	IVA	32	Ge	Germanium	72.59	4	2830	937.4	5.32	$[Ar]3d^{10} 4s^2 4p^2$
4	VA	33	As	Arsenic	74.9216	± 3,5	613*	817	5.72	$[Ar]3d^{10} 4s^2 4p^3$
4	VIA	34	Se	Selenium	78.96	6,4-2	685	217	4.79	$[Ar]3d^{10} 4s^2 4p^4$
4	VIIA	35	Br	Bromine	79.904	± 1,4,5	58	−7.2	3.12	$[Ar]3d^{10} 4s^2 4p^5$
4	INERT GASES	36	Kr	Krypton	83.80	0	−152	−157.3	2.6	$[Ar]3d^{10} 4s^2 4p^6$
5	IA	37	Rb	Rubidium	85.4678	1	688	38.9	1.53	$[Kr]5s^1$
5	IIA	38	Sr	Strontium	87.62	2	1380	768	2.6	$[Kr]5s^2$
5	IIIB	39	Y	Yttrium	88.9059	3	2927	1509	4.47	$[Kr]4d^1 5s^2$
5	IVB	40	Zr	Zirconium	91.22	4	3580	1852	6.49	$[Kr]4d^2 5s^2$
5	VB	41	Nb	Niobium	92.9064	5,3	3300	2415	8.4	$[Kr] 4d^4 5s^1$
5	VIB	42	Mo	Molybdenum	95.94	6,5,4,3,2	5560	2610	10.2	$[Kr]4d^5 5s^1$
5	VIIB	43	Tc	Technetium	(98)	7	—	2200	11.5	$[Kr]4d^5 5s^2$
5	VIII B	44	Ru	Ruthenium	101.07	2,3,4,6,8			12.2	$[Kr]4d^7 5s^1$
5	VIII B	45	Rh	Rhodium	102.9055	2,3,4	4500	1966	12.4	$[Kr]4d^8 5s^1$
5	VIII B	46	Pd	Palladium	106.42	2,4	3980	1552	12.0	$[Kr]4d^{10} 5s^0$
5	IB	47	Ag	Silver	107.868	1	2210	960.8	10.5	$[Kr]4d^{10} 5s^1$
5	IIB	48	Cd	Cadmium	112.41	2	765	320.90	8.65	$[Kr]4d^{10} 5s^2$
5	IIIA	49	In	Indium	114.82	3	2000	156.2	7.31	$[Kr]4d^{10} 5s^2 5p^1$
5	IVA	50	Sn	Tin	118.69	4 2	2270	231.9	7.30	$[Kr]4d^{10} 5s^2 5p^2$
5	VA	51	Sb	Antimony	121.75	± 3,5	1380	630.5	6.62	$[Kr]4d^{10} 5s^2 5p^3$
5	VIA	52	Te	Tellurium	127.60	6,4,-2	989.8	449.5	6.24	$[Kr]4d^{10} 5s^2 5p^4$
5	VIIA	53	I	Iodine	126.9045	± 1,4,5,7	183	113.7	4.94	$[Kr]4d^{10} 5s^2 5p^5$
5	INERT GASES	54	Xe	Xenon	131.29	0	−108.0	−111.9	3.06	$[Kr]4d^{10} 5s^2 5p^6$
6	IA	55	Cs	Cesium	132.9054	1	690	28.7	1.90	$[Xe]6s^1$
6	IIA	56	Ba	Barium	137.33	2	1640	714	3.5	$[Xe]6s^2$
6	IIIB	57	La	Lanthanum	138.9055	3	3470	920	6.17	$[Xe]5d^1 6s^2$
6	IVB	72	Hf	Hafnium	178.49	4	5400	2222	13.1	$[Xe]4f^{14} 5d^2 6s^2$
6	VB	73	Ta	Tantalum	180.9479	5	5425	2996	16.6	$[Xe]4f^{14} 5d^3 6s^2$
6	VIB	74	W	Wolfram	183.85	6,5,4,3,2	5930	3410	19.3	$[Xe]4f^{14} 5d^4 6s^2$
6	VIIB	75	Re	Rhenium	186.207	7,6,4,2,-1	5900	3180	[illegible]	$[Xe]4f^{14} 5d^5 6s^2$
6	VIII B	76	Os	Osmium	190.2	2,3,4,6,8	5500	2700	22.6	$[Xe]4f^{14} 5d^6 6s^2$
6	VIII B	77	Ir	Iridium	192.22	2,3,4,6	5300	2454	22.5	$[Xe]4f^{14} 5d^7 6s^2$
6	VIII B	78	Pt	Platinum	195.08	2,4	4530	1769	21.4	$[Xe]4f^{14} 5d^{10} 6s^0$
6	IB	79	Au	Gold	196.9665	3,1	2970	1063	19.3	$[Xe]4f^{14} 5d^{10} 6s^1$
6	IIB	80	Hg	Mercury	200.59	2,1	357	−38.4	13.6	$[Xe]4f^{14} 5d^{10} 6s^2$
6	IIIA	81	Tl	Thallium	204.383	3,1	1457	303	11.85	$[Xe]4f^{14} 5d^{10} 6s^2 6p^1$
6	IVA	82	Pb	Lead	207.2	4,2	1725	327.4	11.4	$[Xe]4f^{14} 5d^{10} 6s^2 6p^2$
6	VA	83	Bi	Bismuth	208.9804	3,5	1560	271.3	9.8	$[Xe]4f^{14} 5d^{10} 6s^2 6p^3$
6	VIA	84	Po	Polonium	(209)	4,2	—	254	(9.2)	$[Xe]4f^{14} 5d^{10} 6s^2 6p^4$
6	VIIA	85	At	Astatine	(210)	± 1,5,7	—	(302)		$[Xe]4f^{14} 5d^{10} 6s^2 6p^5$
6	INERT GASES	86	Rn	Radon	(222)	0	(−61.8)	(−71)	4.4	$[Xe]4f^{14} 5d^{10} 6s^2 6p^6$
7	IA	87	Fr	Francium	(223)	1	—	(27)	—	$[Rn]7s^1$
7	IIA	88	Ra	Radium	226.0254	2	1737	700	5.0	$[Rn]7s^2$
7	IIIB	89	Ac	Actinium	227.0278	3	—	1050		$[Rn]6d^1 7s^2$
6		58	Ce	Cerium	140.12	3,4	3468	795	6.67	$[Xe]4f^2 5d^0 6s^2$
6		59	Pr	Praseodymium	140.9077	3,4	3127	935	6.77	$[Xe]4f^3 5d^0 6s^2$
6		60	Nd	Neodymium	144.24	3	3027	1024	7.00	$[Xe]4f^4 5d^0 6s^2$
6		61	Pm	Promethium	(145)	3	—	(1027)		$[Xe]4f^5 5d^0 6s^2$
6		62	Sm	Samarium	150.36	3,2	1900	1072	7.54	$[Xe]4f^6 5d^0 6s^2$
6		63	Eu	Europium	151.96	3,2	1439	826	5.26	$[Xe]4f^7 5d^0 6s^2$
6		64	Gd	Gadolinium	157.25	3	3000	1312	7.89	$[Xe]4f^7 5d^1 6s^2$
6		65	Tb	Terbium	158.9254	3,4	2800	1356	8.27	$[Xe]4f^9 5d^0 6s^2$
6		66	Dy	Dyspiosium	162.50	3	2600	1407	8.54	$[Xe]4f^{10} 5d^0 6s^2$
6		67	Ho	Holmium	164.930	3	2600	1461	8.80	$[Xe]4f^{11} 5d^0 6s^2$
6		68	Er	Erbium	167.26	3	2900	1497	9.05	$[Xe]4f^{12} 5d^0 6s^2$
6		69	Tm	Thulium	168.9342	3,2	1727	1545	9.33	$[Xe]4f^{13} 5d^0 6s^2$
6		70	Yb	Ytterbium	173.04	3,2	1427	824	6.98	$[Xe]4f^{14} 5d^0 6s^2$
6		71	Lu	Lutetium	174.967	3	3327	1652	9.84	$[Xe]4f^{14} 5d^1 6s^2$
7		90	Th	Thorium	232.0381	4	3850	1750	11.7	$[Rn]6d^2 7s^2$
7		91	Pa	Protactinium	231.0359	5,4	—	1560	15.4	$[Rn]5f^2 6d^1 7s^2$
7		92	U	Uranium	238.0289	6,5,4,3	3818	1132	19.07	$[Rn]5f^3 6d^1 7s^2$
7		93	Np	Neptunium	237.0482	6,5,4,3	—	637	19	$[Rn]5f^4 6d^0 7s^2$
7		94	Pu	Plutonium	(244)	6,5,4,3	3235	640	—	$[Rn]5f^6 6d^0 7s^2$
7		95	Am	Americium	(243)	6,5,4,3	—	—	11	$[Rn]5f^7 6d^0 7s^2$
7		96	Cm	Curium	(247)		—	—	—	$[Rn]5f^7 6d^1 7s^2$
7		97	Bk	Berkelium	(247)	4,3	—	—	—	$[Rn]5f^7 6d^2 7s^2$
7		98	Cf	Californium	(251)	3	—	—	—	$[Rn]5f^9 6d^1 7s^2$
7		99	Es	Einsteinium	(252)	3	—	—	—	
7		100	Fm	Fermium	(257)	—	—	—	—	
7		101	Md	Mendelevium	258	—	—	—	—	
7		102	No	Nobelium	(259)	—	—	—	—	
7		103	Lr	Lawrencium	(260)	—	—	—	—	

24	Compost - saw dust	1.0	0.7	1.0
25	Compost - farm sweepings	0.4	0.6	0.7
26	Compost sludge	2.5	1.5	0.7
	GREEN MANURES*			
27	Daincha	3.5	0.6	1.2
28	Sunnhemp	2.3	0.5	1.8
29	Wild indigo leaves	3.2	0.3	1.4
30	Glyricidia cuttings	2.9	0.5	2.8
31	Forest leaves	1.2	0.6	0.4
	OILCAKES*			
32	Groundnut cake	7.6	1.3	#
33	Castor cake	5.3	1.6	#
34	Neem cake	4.7	1.9	#
35	Pungam cake	4.2	0.9	#
36	Gingelly cake	6.1	2.4	#
37	Safflower cake	5.5	1.0	#
38	Coconut cake	3.5	1.4	#

* The nutrient content given for manures, green leaf manures and oil cakes are only approximate and will vary depending upon source, method of preparation and other factors.

Normally the amount of potassium is not estimated in these as it is very low.

INDEX